RADIATION
PROTECTIVE
FOODS
A MENU FOR THE NUCLEAR AGE

RADIATION PROTECTIVE FOODS

A MENU FOR THE NUCLEAR AGE

SARA SHANNON

authorHOUSE®

AuthorHouse™
1663 Liberty Drive
Bloomington, IN 47403
www.authorhouse.com
Phone: 1-800-839-8640

The publisher does not advocate the use of any particular diet and exercise program, but believes the information presented in this book should be available to the public.
Because there is always some risk involved, the author and publisher are not responsible for any adverse effects or consequences resulting from the use of any of the suggestions, preparations or procedures in this book. Please do not use the book if you are unwilling to assume the risk. Feel free to consult a physician or other qualified health professional. It is a sign of wisdom, not cowardice, to seek a second or third opinion.

Published by AuthorHouse   05/11/2012

ISBN: 978-1-4670-3573-6 (sc)
ISBN: 978-1-4670-3574-3 (e)

Library of Congress Control Number: 2011916605

Any people depicted in stock imagery provided by Thinkstock are models, and such images are being used for illustrative purposes only.
Certain stock imagery © Thinkstock.

This book is printed on acid-free paper.

Because of the dynamic nature of the Internet, any web addresses or links contained in this book may have changed since publication and may no longer be valid. The views expressed in this work are solely those of the author and do not necessarily reflect the views of the publisher, and the publisher hereby disclaims any responsibility for them.

THIS BOOK IS DEDICATED TO THAT TIME

WHEN THE SACREDNESS OF THE EARTH

WILL BE RECOGNIZED BY EVERYONE.

CONTENTS

Acknowledgments...1
Preface 2012...3
Preface 1987...15
Introduction 1987...19

1. Radiation in the Environment 27
2. How Radiation Damages Us..66
3. Survival of the Fittest..92
4. Your Body's Radioprotective Systems 106
5. The Foods to Avoid...136
6. A Menu for the Nuclear Age...153
7. Supportive Supplements 205
8. Recipes..226

Afterword..249
Appendix 2012..251
References...264
Index 315

Acknowledgments

I would like to express my gratitude to all those who have contributed to this book, both directly and indirectly. Since I first began writing this book a few years ago, I have spoken with hundreds of people who offered their expertise in a variety of disciplines. It is impossible to thank each by name, but their assistance has been invaluable.

In particular, I appreciate the efforts of Dr. Ernest J. Sternglass, Professor Emeritus of Radiological Physics, School of Medicine, University of Pittsburgh, for his painstaking review of the manuscript and for taking the time on several different occasions to discuss his research in nontechnical terms. My appreciation also goes to Dr. Rosalie Bertell, Director of Research at the International Institute of Concern for Public Health, who reviewed the manuscript in-depth and offered many pertinent suggestions and insights into the problems of radiation and health. In addition, thanks are due to Dr. Martin Levine, Professor of Chemistry at City University of New York, for his assistance with the material relating to chemistry in the first three chapters.

PREFACE
2012

This book was first published in 1986 with the title *Diet for the Atomic Age*. My aim at that time was to propose a radio-protective way of eating. This is still valid and even more important now. We can take steps to protect ourselves through the use of the food we eat. Yet, there is a limit. Time goes on and cumulative levels of all pollutants increase.

In the energy domain we must put our efforts toward reducing and stopping a prime source of radiation poisoning: the use of nuclear power. We now have over 400 nuclear reactors worldwide. The entire range of nuclear technologies, including uranium mining, reprocessing radioactive waste, the military use of depleted uranium, and other related activities increase the radioactive load. We must encourage the implementation of clean renewable sources of energy such as sun, geothermal, wind and water as replacements for nuclear power.

The March 2011 nuclear catastrophe in Japan and previous catastrophes in Russia and the United States are lessons to us that nuclear power is not the safe energy option that many people would have us believe.

Disinformation is rampant. For example, the media in Japan and worldwide were only given the partial information TEFCO and the government saw fit to give them about the impact of the Fukushima accident. The truth might be too damaging to the nuclear industry. In fact, three weeks after the disaster 3 reactors were in various stages of collapse having melted through the core with molten fuel going into the ground. Fuel rods were exposed and desperate attempts were made to keep them from exploding by cooling them with sea water, and then allowing this radioactive water to seep into the Pacific Ocean. More than 2.6 million gallons have leaked or been released into the sea. (Bloomberg News 5/18/11 "TEPCO Misleading Public over Nuclear Crisis").

Oceanographers know that the current sweeps from Japan to the central Pacific Ocean and then to the west coast of the United States and beyond. The several Fukushima explosions sent radioactive materials into the global air currents crossing the American continent. The Environmental Protection Agency (EPA) does not report from their radiation monitors, and we are allowed the impression that there is not much to be concerned about.

The accident in Japan is a global disaster and will have profound effects on all forms of life all over the world.

Dr. Christopher Busby, scientific advisor of the European Committee on Radiation Risk and to the Low Level Radiation Campaign (LLRC) stated in a May 17 2011 RT news interview that Fukushima "is out of control. It is a global problem. The fissioning needs to be contained." Michio Kaku, a nuclear physicist, in an interview April 13, 2011 (Democracy Now) responding to a question about Fukushima said, "They are making it up as they go along, and we are the Guinea Pigs."

The Nuclear Regulatory Commission (NRC), whose job it is to monitor the safe operation of our 104 nuclear power plants, has proven to be lax regarding the nuclear industry, ignoring blatant safety violations, and rubber stamping 20 year operating extensions for more than half the aging plants which were only designed to last 40 years. These plants have already been pressed into service well past their expiration dates, and many of them have degrading cement towers and other problems.

Arnie Gundersen, a former nuclear-engineer-turned-whistle-blower tells us:

> "The NRC has consistently put industry profits above public safety. Consequently we have a dozen Fukushimas waiting to happen in America."

The late professor John Gofman PhD MD was a senior figure in the Atomic Energy Commission (AEC) in the 1960's. He was the director of a government lab doing research on the effects of radiation on human health. Dr. Gofman found that radiation did more damage, and in different ways, than had been suspected. When his data was buried, he resigned. With his specialized expertise and his integrity intact, and having been the victim of a campaign to launder the science to fit the goals of the industry, he concluded:

> "The nuclear industry is waging a war against humanity."

Dr. Christopher Busby who has conducted extensive research on the health effects of radiation (you can see his various videos on YouTube) expands John Gofman's statement in an article titled "Deconstructing Nuclear Experts" (3/28/11 *Counterpunch*)

> "The war has now entered an end game which will decide the survival of the human race. Not from sudden nuclear war, but from the on-going and incremental nuclear war which began with the releases to the bioshpere in the 60's of all the atmospheric test fallout, and which has continued inexorably since then through Windscale, 3-Mile Island, Chernobyl, Hanford, Sellafield, La Haque, Iraq and now Fukushima, accompanied by parallel increases in cancer rates and fertility loss to the human race."

Indeed, in the 1960's the Atomic Energy Commission knew very well what they were releasing, what war they were waging. A 1966 report by the AEC acknowledged the damaging potential of radiation saying "there is no threshold on the genetic effect of radiation." This report, "The Genetic Effects of Radiation" said:

> "If the number of those affected is increased, there would come a crucial point or threshold, where the slack could no longer be taken up (by those not affected). The genetic load might increase to the point where the species *as a whole* would degenerate and fade toward extinction—a sort of 'racial radiation sickness.'" (italics added)

These acknowledgments aside, the AEC went ahead and set a number amount on allowable radiation exposure from nuclear power plants' normal operation without warning that this amount causes some of the population certain health damage. Quoting John Gofman again:

> "Licensing a nuclear power plant is, in my view, licensing random premeditated murder."

As a medical doctor and honest researcher in radiation health John Gofman refers to the known attack on health from the every day operation of nuclear power plants.

Nuclear power can be disastrous in the case of an accident. However, what is not well understood by the public is that nuclear power is damaging in its normal, routine operation. Nuclear plants regularly release radioactive material as a part of their every-day operation. These routine emissions are described as low-level and safe, but in reality they are a continuing assault on human health and the environment.

This health damage includes infertility, mental retardation, cancer, infectious diseases, asthma, diabetes and arthritis. Exposure to radioactivity causes a general weakening of the immune systems of individuals and damage to the genetic health of the species as a whole.

It has been the common wisdom that the effects of low-level doses of radiation are negligible and that they cannot be studied directly.

Substantial research rejects this approach. A number of independent researchers have demonstrated that long-term, relatively low levels of radiation may wreak up to 1000 times more biological havoc than the currently accepted "risk levels". These "risk levels" are used as standards for setting "safe" levels of exposure and as reference points for decisions about licensing and operating nuclear plants.

See chapter 2 for a discussion of the research relating to radiation health damage and how the sobering facts were slow to filter through to the public.

See chapter 3 for an understanding of the "healthy survival theory" and the principle of "selective uptake". There is a wide range of sensitivity to radiation and those who are more healthy are more fortified against it. This is the basic concept and foundation for the way of eating which will provide a radiation shield.

Chapter 4 explains the body's defense systems and the critical importance of detoxing the body of other toxins so that it is freed up to combat radioactivity.

Chapters 5, 6, and 7 expand on foods and supplements, followed by some sample recipes.

This book also has a unique collection of references. See the new appendix titled *Hazards of Low Level Radioactivity* which discusses effects of radiation which are not specific health damage.

Since Japan's disaster, Spain, China, Germany and other countries have decided to review the safety of their nuclear plants. The German government agreed to phase out all nuclear power by 2022. In contrast, the U.S. plans to go forward with the $54.5 billion budget in loan guarantees for new nuclear plants and to extend the operating life of over 60 existing plants. Companies continue their efforts to sell nuclear technology abroad.

Commenting on the US administration's proposed $54.5 billion in Federal loan guarantees to fund new nuclear reactors, Robert

Alvarez, an Institute for Policy Studies Senior Scholar who was Policy Advisor to the Energy Department from 1993 to 1999, stated:

> "Because of skyrocketing costs these loans might pay for
> five reactors, and merely expand the nations electrical
> supply by less than 1 percent."

("Nukes Aren't the Answer" <u>CommonDreams.org</u> Feb. 15, 2010)

That means we would be paying over $54 billion for 1 percent of our electric needs!

Another critical fact to consider is that the long touted repository for nuclear waste at Yucca Mountain in Nevada was cancelled in the spring of 2011; this was after spending $9 billion to try to develop it. There is now no long term plan for radiation waste storage. Used fuel rods are being "double-racked", that is they are being packed dangerously tightly in the cooling pools near the power plants because there is no place to send them for long term storage. We know from Fukushima that this is an incredibly dangerous situation.

There is a lot to be done. In this era after Fukushima we can start by empowering ourselves.

Foods can promote optimal health and create the most protective internal environment in living species. Eating 'low on the food chain' and utilizing the important principle of 'selective uptake' while avoiding genetically modified foods and food grown with pesticides will help build high energy cells. These cells are better able to resist radiation as well as other environmental assaults.

This book describes the crucial problem of nuclear power and offers a way to shield yourself from ongoing levels of radiation by the use of foods with protective properties, all based on medical and scientific data.

We are at a crossroads. This is one of the most serious challenges humanity has faced.

We must phase out death-dealing nuclear power and phase in clean, renewable energy sources. With the global air currents carrying depleted uranium from munitions used in Iraq and other military invasions, compounded by the radioactivity from the Fukushima catastrophe, on top of the releases from the everyday normal operation of the worlds 400 plus nuclear power plants, we have no time to waste.

During this profoundly pathological time, there is opportunity for change. It is up to us!

Everyone is encouraged to develop their own sources of food, whether through participation in community-supported agriculture (CSA), supporting local farmers, or by farming and gardening themselves. People can explore biodynamic agriculture, greenhouses and water filtration systems.

* * * * * * * * * * * *

This issue concerns all of us. Although many activists are marginalized, some respected scientists working in the field have come to strong conclusions about the dangers of nuclear power.

Nuclear power is the greatest public health scandal in human history.

Christopher Busby PhD

> Dr. Chris Busby, a British scientist respected for his work on the negative health effects of low-dose radiation, has been focused recently on the health effects of ingested depleted uranium particles. He is the director of the Green Audit Company and scientific advisor to the Low Level Radiation campaign (LLRC) which he set up in 1995. He is a professor at the University of Ulster and author of many research articles and two books. (See *Wings of Death: Nuclear Pollution and Human Health*)

● ● ●

Should the public discover the true health costs of nuclear pollution, a cry would rise from all parts of the world and people would refuse to cooperate passively with their own death.

Dr. Rosalie Bertell

Dr. Rosalie Bertell, founder of the International Institute of Concern for Public Health, has a doctorate in biometrics and five honorary doctorates. She is world-famous in the field of environmental health since 1969, and led the Bhopal and Chernobyl Medical Commissions. She is the recipient of the Right Livelihood Award, the alternative Nobel Prize, and the author of many articles and two books. (See *No Immediate Danger: Prognosis for a Radioactive Earth"*)

● ● ●

There is no safe amount of radiation. Even small amounts do harm.

Dr. Linus Pauling

Dr. Linus Pauling was a chemist and peace activist. He was recipient of the Nobel Prize in Chemistry in 1954: He revolutionized chemistry with his application of quantum physics to the study of chemistry. In 1957 he petitioned the United Nations to end the atmospheric testing of nuclear weapons, emphasizing the hazards of radiation. He won the Nobel Peace Prize in 1962.

● ● ●

The nuclear industry is waging a war against humanity . . . no right is closer to the constitutional guarantees than that of a livable environment.

Dr. John Gofman MD, PhD

Dr. John Gofman was a medical doctor with a PhD in nuclear-physical chemistry. He was the Associate

Director of the Lawrence Livermore Radiation Lab in Livermore California from 1963 to 1969. He was a professor of radio-biology for over 20 years. Gofman was a senior figure in the Atomic Energy Commission, a director of a research laboratory focussing on health and radiation. When his data didn't conform to his employers expectations, he was forced to resign.

●   ●   ●

Websites:

A list of nuclear plants in the USA with their locations and details such as the amount of high-level radioactive waste on site, problems and leaks, and "worst case" projections of casualties and property damage:
www.animatedsoftware.com/environm/no_nukes/nukelist1.htm

A list of 200 nuclear related books and videos:
www.animatedsoftware.com/environm/no_nukes/mybooks.htm

A national information center about nuclear power and sustainable energy issues:
www.nirs.org

A map of the USA depicting radiation levels updated frequently:
www.radiationnetwork.com

About the catastrophe in Japan:
www.consciousbeingalliance.com/2011/03/japans-catastrophe-nuclear-power-cover-up
www.fairewinds.com

Dr. Chris Busby and Leuren Moret both have videos on YouTube.

About Health Effects of Low Level Radiation:
www.nuclearreader.info
www.grassrootspeace.org
www.llrc.org
www.nirs.org
www.beyondnuclear.org
www.antennanl/wise
www.llcph.org
www.radiation.org

About Nikola Tesla: a video "Nikola Tesla: The Forgotten Wizard of Free Energy"
www.forbiddenknowledge.com/videos/alternative-energy/nikola-teslathe-forgotten wizard-of-free-energy or search on YouTube

About biodynamic agriculture:
"Grow a Garden and be Self-Sufficient" by E. Pfeiffer and E. Riese
"Biodynamic Greenhouse Management" by Heinz Grotzke

A website clarifying the spectrum of nuclear power:
www.nuclearreader.info

DISCLAIMER

The health conclusions in this book are based on extensive medical and scientific research. However, nothing herein constitutes personal medical advice. By reading this book there is an implicit agreement that actions you may take are your sole responsibility.

This book was written with the idea in mind that the responsibility to preserve and build our health belongs to each of us and the more well-documented information we have, the more healthy we can be.

Preface
1987

My awareness of the health problems developing as a result of exposure to radioactive fallout and radioactive leaks from nuclear plants developed slowly. Even though I was working as a nutritionist and keeping up with new research, I rarely carne upon articles on the subject.

At a conference I attended in 1977, a famous author of health books was asked what one vitamin he would take if he was going to be stranded on a deserted island. I thought he would probably say vitamin C because it has the most overall fortifying and protective action, but I was wrong. His answer was "Kyolic." This is a special concentration of garlic. "I would choose this," he explained, "because it detoxifies radiation." I was impressed by this thought and by the idea that the foods we eat could counteract the hazardous effects of radiation.

At the time of the near-disaster at Pennsylvania's Three Mile Island nuclear power plant in March 1979, I was in New York City and I got packed and ready to get out of the area. The radio reports that the situation was "stable" didn't convince me at all. My instinct spoke to me. I kept my son horne from school. When I called to say he wouldn't be back that Monday, I told the assistant principal why. He replied that he'd heard that President Carter was going to the accident site and that the problem was stable. I was amazed that he, one who is responsible for children, would be so cool. I asked if many other mothers had called and he said not one, and definitely let me know I was being extreme-if not a kook.

A few months later the *East West Journal* had an article entitled "Natural Ways to Survive a Meltdown." As a nutritionist, mother and ordinary citizen, I read this article very intently. It spoke of rice and seaweed and stated that food is our best protection from radiation, listing which foods to eat and which to avoid.

In the middle of August 1982, I read an article in the *New York Times* about Utah sheep farmers who had sued the government in 1956 after losing over 4,000 sheep to what they said was radiation poisoning from aboveground bomb tests nearby. A Federal judge reversed the initial decision that radiation had nothing to do with the death of the livestock, ruled that the government had engaged in fraud and deception to prove its point, and ordered a new trial. (As of this writing, however, the trial order has been overturned and an appeal from the ranchers has been denied.) One of the ranchers quoted in the magazine article said, "I'm sure they felt if this whole can of worms was opened up beginning with sheep, the people would rise up against the testing."

A few days later, walking through Manhattan's Central Park, I came to a peace rally on the anniversary of the first use of atomic weapons-the bombing of Hiroshima. There were various booths staffed with volunteers handing out pamphlets. Very few people were at the rally, althought it was a bright yet cool Saturday. I came to one table that had a large picture book set on a book stand. Leafing through it, I saw horrifying photos documenting the death and destruction at Hiroshima days after the bomb. The agony caused by radiation sickness was apparent everywhere; the bomb victims suffered extensive flash burns and disfigurement. The woman at the table explained to me that sale and distribution of this book had been prohibited in the United States. She had gotten this single copy from a group of Japanese visitors here.

These pictures had a powerful effect on me. I called a few nuclear protest groups to see if they had any information on how to protect ourselves from radiation. They all seemed to be focused on the question of nukes or no nukes. A typical response- from the Union of Concerned Scientists was, "We deal mainly with alternatives to nuclear energy and warfare. We don't deal with anything concerning health and diet. In fact, there is no way to protect yourself."

What kind of information about nuclear hazards is readily available to concerned citizens? Can we protect ourselves? I browsed in bookstores to find out. In many stores, Jonathan Schell's *The Fate of the Earth* was featured on a front shelf.The author proposes we "lay down our arms" and aim for a peaceful discussion of problems. The book *Killing Our Own*, by Harvey Wasserman and Norman Soloman, was on a back shelf-more controversial because its topic is the danger of radioactive fallout. I remembered a small book I'd once read, *Are You Radioactive?*, by Linda Clark, and discovered that it is out of print. *Nuclear Madness* by Dr. Helen Caldicott was here and there in the smaller bookstores. It's an adamant and convincing indictment of the whole nuclear industry, but again mainly concerned with avoiding war.

Checking through a local bookstore I found *Secret Fallout* by Dr. Ernest Sternglass, a professor of radiation physics who talks emphatically about the cumulative and insidious effects of radiation on our health-ranging from weakening of the immune system and increased genetic mutations to cancer, tumors and a host of diseases. But even this book did not give advice on what to do about this or how we can protect ourselves.

The only available source of this information was the *East West Journal*, a macrobiotic-oriented magazine sold through health food stores. In the last few years, *EWJ* has had several articles on how to use diet to protect and cleanse oneself from fallout. I obtained these back issues and talked to an editor at the magazine. He told me, "There is no conclusive evidence, mostly conjecture. I really can't say.... Read the article."

I read everything I could. I spent many afternoons in libraries searching for information about the connection between health and radiation. I came to realize that nuclear fallout and power plant emissions are not the only causes of radioactive contamination- the transportation and storage of both high-level and low-level radioactive wastes are hazardous as well. Moreover, I became aware of health risks from *non-ionizing radiation* which comes from sources like microwave ovens and television sets. Such radiation exposure is less intense but more pervasive than exposure to *radioactivity*. Both can cause problems.

My research led me to some fascinating articles; I learned that vitamins and minerals, the acid-alkali balance and one's type of diet can all have a *radioprotective* effect. There were a few books with titles like *Chemical Protection Against Ionizing Radiation*, which told about chemical chelating agents that are usually very toxic in themselves, but can help the body to remove radiation.

Although low-level exposure to the various forms of radiation is something that everyone in our society encounters on a daily basis, I *couldn't* understand the paucity of information about how we *can* protect our health. All in all, I found data confirming that we can protect ourselves from radiation and information on how we can do so scattered and sparse.

I decided to keep looking. This book, *Diet for the Atomic Age*, tells what I found.

Note: This new edition is re-titled Radiation Protective Foods

Introduction
1987

by Dr. Ernest J. Sternglass

In writing this book, Sara Shannon has done a major service to the future health and survival of our species on this globe. She has brought together the evidence that radioactive substances appear to be the single most powerful of all the various toxic materials in our diet, and that it is to a large extent in our power to protect ourselves and our children from their harmful effects. What I would like to do here is to briefly describe what it is that we did not understand about the danger of radioactivity in the diet, why it took so long for us to arrive at this understanding, and what can be done to minimize the harm of radioactivity already in our food and water.

When the first atomic bomb was detonated in Alamogordo, New Mexico in July of 1945, little was known about the potential hazard of radioactive chemicals created in the fission of uranium or plutonium entering the human diet. Fission itself had only been discovered six years earlier, and only extremely small amounts of the newly formed radioactive elements had been available to scientists for study. All the effort was being concentrated on building a bomb before German or Japanese scientists could do so, leaving little time for biological studies of the effect of radioactive fallout, especially at the low levels likely to be encountered far from the point of the actual explosion. And whatever knowledge of biological effects was being obtained at government laboratories was classified by the military, since the very existence of a program to develop a nuclear weapon was one of the most closely guarded secrets of the entire war.

Most of our knowledge of the hazard of radiation until that time had come from the medical uses of radiation, and this involved entirely the effects of X-rays and gamma rays, which were discovered between 1895 and 1896. All this radiation came from external sources such as X-ray machines or radium capsules, not from radioactive materials inside the human body. Only a handful of workers in factories using radium mixed with luminescent chemicals to paint the dials on instrument panels had ingested small amounts of the radium accidentally, leading to bone cancers many years later. And although large amounts of radiation needed to kill cancer cells in the course of radiation therapy were known to produce serious damage to healthy tissue also, there was no evidence at all that radiation at the low levels of diagnostic X-rays some one thousand to ten thousand times weaker than therapeutic doses had any damaging effects at all. Also healthy tissue in the human body irradiated in the course of cancer treatment had a remarkable capability to recover when the radiation treatment was spread out over a period of weeks or months, a fact which made radiation therapy possible in the first place. Thus, a half century of medical experience seemed to indicate that there was a safe threshold below which no observable damage to human health would occur, a threshold that appeared to lie at doses equivalent to perhaps a thousand to ten thousand chest X-rays.

Furthermore, when it was discovered in 1927 that X-rays do produce damage to reproductive cells in animals, leading to birth defects in later generations, the dose required to do so was found to be very large, or about ten thousand times that encountered in a chest X-ray. Compared to the dose we receive from natural radiation sources in our environment during the course of a whole year, the dose needed to double the small number of genetic mutations that normally occurred was some one thousand times greater.

All the experience gained from the medical uses of radiation and laboratory studies of mutations therefore suggested that small amounts of fallout produced by distant atomic detonations at doses well below those from natural background sources would have no detectable effects on human health. In view of this evidence, a decision was made to carry out a series of nuclear tests in the Pacific starting in 1946, and later in Nevada beginning in early 1951. The cold war had started, and the Korean War had led to the fear that without the use of nuclear weapons it might not be possible to stop the large armies of the Chinese and Russian war machines.

In this climate of fear, the political and military leadership training troops for nuclear warfare was not anxious to see research on the possibility that the inhalation or ingestion of very small amounts of radioactive materials from bomb fallout might be more dangerous than external radiation on the ground. A few scientists had begun to realize that the dose to certain organs in the human body as a result of inhalation would be much larger than that produced by the gamma rays from the radioactive material on the grass or soil if a cloud of radioactive fission products were to pass by. And other studies by a few concerned scientists had begun to show that eating the contaminated vegetables or drinking the milk produced by cows grazing on the contaminated grass would also lead to much larger internal doses to certain organs than the external radiation, in some cases ten to a hundred times larger. But the political need to continue bomb testing during the height of the confrontation with the Soviet Union over Berlin and Cuba persuaded the top officials in the Pentagon, the Atomic Energy Commission and the White House that research in this area should be limited to laboratory studies, and that no large-scale epidemiological studies of human populations exposed to fallout in their diet should be further pursued or published. Political leaders feared that any published information indicating serious effects on human health would weaken the deterrent value of nuclear weapons on which the security of the nation rested.

As a result, it did not become widely known in the scientific community, in the medical community or among the general public at large that somehow, low doses of radiation due to chemical species resulting from the fission process were producing leukemia and cancer cases at a rate hundreds to thousands of times greater than had been calculated on the basis of our experience with high doses at Hiroshima and Nagasaki, or our experience with medical radiation and vast numbers of animal studies in the laboratory in over a half-century of research. And although very disturbing upward changes in total mortality and infant mortality were being discovered all over the world during the late 1950s and early '60s, there was no known scientific basis for the possibility that they might be related to the small doses from fallout.

Also, by the end of the 1950s, the first commercial nuclear reactors had been completed, and there was great hope among scientists and engineers that the peaceful use of nuclear fission would bring great benefits to society, outweighing the negative aspects produced by the development of the bombs, and thereby atoning for their use at Hiroshima and Nagasaki. By the early 1960s, large numbers of giant nuclear reactors were on the drawing boards, and enormous investments were being made to replace dirty coal plants with clean nuclear power, promising an end to lung diseases and dependence on foreign oil. The signing of a treaty between the U.S., Great Britain and the U.S.S.R. in 1963 to end bomb testing in the atmosphere had ended widespread concern among many scientists and members of the public about fallout, and most scientists returned to their normal pursuits, so that the problem of low-level radiation effects was soon forgotten.

The realization that small amounts of radiation could be much more harmful than had been suspected came quite unexpectedly in the course of a study begun in the late 1950s by Dr. Alice Stewart in England. While trying to understand why leukemia among young children had risen so sharply after World War II, Dr. Stewart discovered that when women were exposed to just a few diagnostic X-rays during pregnancy, their children developed leukemia twice as often as normally expected. Not until her findings had been substantiated in a separate study by Dr. Brian MacMahon at Harvard, which was published in 1962, was it possible for radiologists and obstetricians to believe, in the face of half a century of medical experience with diagnostic X-rays that seemed to show no ill effects, that leukemia and cancer could be induced in humans at less than one-hundredth the dose it took to double the risk of these diseases in adults.

But the situation was destined to become much worse. Within a few more years, Dr. Stewart had gathered enough data to show that for irradiation in the first three months of the mother's pregnancy, the child's risk of developing leukemia in childhood was ten times larger than for irradiation at the end of pregnancy. Thus, the equivalent of a single year of natural background radiation, widely believed to be harmless, now seemed to double the risk of many different types of cancer in a human population, suddenly explaining why small amounts of fallout could indeed produce serious illness, especially in the young whose natural immune defenses are least effective.

The discovery of a thousand fold greater sensitivity in the early human embryo and fetus now could explain why infant mortality due to all causes combined could have been affected by fallout from nuclear testing, especially when combined with the ten-to-one-hundred-times-greater dose to various critical hormone-producing organs, such as the thyroid, the pituitary and the gonads. Strontium-90, which is created in large quantities during the fission process and goes mainly to the bone, irradiates the bone marrow where the cells of the immune system develop, explaining why so many more infants died of infections during the period of bomb testing than expected. The yttrium-90 formed in the radioactive decay of strontium-90 goes to the pituitary, the ovaries and the thymus, all of which are involved in regulating the immune and reproductive systems controlling the birth process and the timing of the onset of labor, thus helping to explain the epidemic of spontaneous miscarriages and premature deliveries that began during the 1950s. And radioactive iodine-131, produced with even greater intensity in the fission process, goes to the thyroid of the fetus where it concentrates a hundred times more than in the thyroid of an adult, lowering the production of a key hormone that affects the fetus' growth and the development of all its organs. These affected organs include the critical lung and brain of the newborn infant, explaining the epidemic of underweight births accompanied by breathing problems, and the increased incidence of brain damage, dyslexia, and subtle forms of learning disabilities that began during the years of nuclear testing.

But the effects of fallout were not confined to the newborn, as data on adult cancer and total mortality paralleled the changes in infant mortality with a time-lag of a few years. The reason for this did not become clear until 1972, when a Canadian physician by the name of Abram Petkau discovered that cell membranes, such as those of the white blood cells involved in the immune system defenses of the body, were damaged much more readily by protracted low-level exposures to radiation than by a brief exposure at the same total dose. He discovered that at the low doses of fallout radiation, the dominant damage was not produced by direct hits of radiation on the DNA in the genes, but by the production of unstable chemicals called free radicals that damaged cell membranes. And this process was a thousand times more efficient when the radiation acted over long periods of time than when given in a brief flash as in an X-ray or by the detonation of a nuclear bomb.

Suddenly it became clear that we had been grossly misled by the absence of serious side effects of diagnostic X-rays in adults into believing that small amounts of radiation from fallout are harmless. Rather, it became clear that for radiation acting slowly over periods of weeks, months or years the damage to human health is actually a thousand to ten thousand times more serious than a short X-ray exposure of the same dose, especially when the large concentration of radioactive chemicals in the diet and the critical organs of the body affecting our immune defenses is taken into consideration.

And since the permissible discharges from commercial nuclear reactors and the allowable concentrations of radioactive materials in the diet were set in the 1950s to permit nuclear weapons tests on the basis of our experience with the brief, intense exposures experienced by the bomb survivors and our knowledge of the effects of short medical exposures for adults, the stage was set for a great tragedy. As more and more nuclear reactors near our cities released fission products into the air and drinking water in the course of normal operations or accidents such as Three Mile Island, mortality rates started to rise again, just as they did at the height of nuclear testing.

However, as a result of these new discoveries, it is now becoming clear that we are not all doomed to die prematurely of cancer, heart disease and infections, or to see our children born with an ever increasing risk of birth defects as is now happening in many areas of the United States. Indeed, this trend has already ended in those states where there are no large nuclear reactors located in or near the border, no bomb production facilities and no nuclear test sites. Total and infant mortality rates have in fact declined sharply in areas such as Wyoming and New Hampshire since the height of nuclear fallout when the rates were as high as in the rest of the United States. In these areas, the mortality rates have now approached the lowest in the world, such as those achieved in Iceland and Denmark, declining steadily since the end of atmospheric bomb tests. That radioactive materials are the dominant factor is further supported by the fact that the people in Wyoming and New Hampshire continue to smoke cigarettes, generate electricity with coal, oil and natural gas, use pesticides and herbicides in agriculture, and have contact with many ordinary chemicals in their workplaces and in their daily lives.

If mortality rates can decline sharply in some areas of our nation, they can be decreased anywhere. It is in our power to decide not to produce plutonium for nuclear weapons or generate electricity with nuclear fission reactors, as the people have decided in countries such as Austria, Denmark and Norway. And as this book shows, we can in the meantime greatly reduce the risk to our health and that of the newborn through our diet, since this is the principal path through which radioactivity enters our body. We can greatly lower the intake of radioactive elements such as strontium-90 relative to the chemically similar calcium which the body needs by our choice of food and drink. We can use foods that actually draw out some of the most toxic radioactive elements from our body by means of kelp and other sea-vegetables widely used in the Japanese and Icelandic diets. And we can greatly strengthen our immune defenses in many ways against the destructive action of free radicals as the author shows. Thus, there is hope that we can achieve a healthier future for ourselves and our children even in the Atomic Age, provided that we are willing to learn from our tragic mistakes and act upon our hard-won knowledge.

—Ernest J. Sternglass
Professor Emeritus of Radiological Physics
School of Medicine, University of Pittsburgh

1

Radiation in the Environment

Everyone is concerned about the danger of nuclear holocaust. What I am addressing is the danger of radiation in the environment *now*. Even if we did succeed tomorrow in stopping all nuclear weapons production and closing all nuclear power plants, we would nevertheless be faced with the monumental task of cleaning up tons of radioactive waste that have accumulated over the years—from atmospheric and underground bomb testing, waste storage sites, and research and industry. The world's nuclear plants will be sources of radioactive contamination for thousands of years after their useful life of thirty years.

Men and women lived in equilibrium with natural low-level radiation in the environment up until the last years of the nineteenth century, when x-rays were discovered. Although it was observed that this form of radiation could have a devastating effect, the early use of x-ray machines went unchecked in medicine. Very little research was done on the negative medical impact of x-rays until the mid-1940s. Much has been learned since then, but not until the early 1970s was the full import understood. In addition to x-rays, several other sources of low-level radiation have been developed in this century. Now there is a new fact of existence. A risk of damage to our health from low-level radiation does exist.

I am sure we are all very sick of hearing *about* pollution, but the fact is that we are getting sick *from* pollution—particularly radiation pollution. The role played by low levels of radiation in the human body has attracted an increasing amount of attention as the use of nuclear power has increased throughout the world. Its potential effects include several types of cancer, leukemia, harmful genetic mutations, mental impairment and fetal and infant deaths. Protection against ionizing radiation becomes more important not only as we become more dependent on nuclear energy, but also as nuclear medicine and technology advance and the nuclear arms race escalates.

It is lucky for us that there are ways to prevent the assimilation of radioactive substances by the human body, and ways to get rid of them once they have been taken up by the body. Foods that have particular protective properties are the secret to helping us to live with radiation. Over two thousand years ago, Hippocrates summed up what's largely been forgotten—"Let your medicine be your food, and your food be your medicine." It sounds too simple. But it is true.

The fact that we *can* protect ourselves from the health-destroying effects of low-level radiation is not generally known. No previous synthesis of the information from the point of view of nutritional status has been made. The Nuclear Age Diet, which is based on traditional whole foods, grew out of my research over the last several years and can help to make you as radiation-resistant as possible.

WHAT IS RADIATION?

Radiation is the emission and transmission of energy. Light and microwaves are examples of radiation. In this book, we will focus on the type of radiation that is known as radioactivity. Radioactivity—whether from the natural background or man-made sources—comes from an instability in the atom. The atom is the fundamental building block of all matter; the glue that holds its parts together is electromagnetic attraction. The center of the atom is known as the nucleus. It is composed of particles called protons and neutrons, and has a positive charge. When the atom is electrically neutral, this positive charge balances the negative charge of the electrons orbiting around it.

The basic elements are defined according to the number of protons contained in the atomic nucleus. The number of protons remains constant for each element. For all of the elements, however, there are possible variations in the number of neutrons. These varying forms of an element are known as *isotopes.* Some isotopes are common, while others are quite rare. Some are stable, but most arc not. Unstable isotopes are said to be *radioactive.*

The instability of the atomic nucleus is a complex phenomenon. Nevertheless, the ratio of neutrons to protons plays an important role. Many different isotopes contain neutron-proton ratios that are inherently unstable. Such unstable isotopes (sometimes referred to as *radioisotopes* or *radionuclides)* release energy and particles to achieve stability. This release of energy and particles is known as *radioactivity.* The nucleus disintegrates (decays) and gives off particles and electromagnetic waves—that is, *energy*—until it reaches a stable state.

The energy released by the atom is called *ionizing radiation* when it is strong enough to remove electrons from other atoms or molecules with which it comes into contact; this is how it damages living tissue. Radioactive particles get into the living organism via air, water or food, and destroy tissues and organs, cells and genes.

Non-ionizing radiation is not strong enough to dislodge electrons from their orbits. This form of energy is all around us in various forms, including visible light, radio waves and microwaves. The electromagnetic spectrum (see next page) shows the relationship between the various types of radiation.

Energy from the atom can appear as motion, light or heat. For example, energy in the form of heat is utilized in power plants to make steam and generate electricity. The atom expresses its energy in the form of motion in x-rays and radio waves. And radium, a naturally occurring radioactive substance, gives off light. Starting on the eve of World War I and continuing until the early 1930s, radium was used to paint luminous watch dials and instrument panels. In 1913, about 8,000 dials were produced; by the year 1919, over 2 million were turned out annually. The women who painted the watch dials were encouraged to lick the paintbrushes so that the painted lines would-be finer. As the years went by, a significant number of these workers developed cancer and died. Nevertheless, some radioactive materials are still used in the manufacture of products with luminous displays.

In many respects, radiation is the greatest contaminant in the world. It cannot be seen, felt or heard. It is tasteless and odorless. It is in our food and in the air; it is in our blood and in our bones and can remain in our ashes to go on to contaminate someone else.

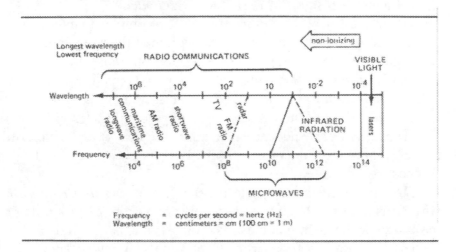

The Electromagnetic Spectrum

Half-Life and Radioactive Life

As we have seen, when an unstable atom decays, it loses particles from its nucleus. The *half-life* of a radioactive substance is the time it takes for half of it to decay. The half-life can range from a fraction of a second to many thousands of years. The important thing to recognize is that when scientists say strontium-90, for example, has a half-life of 28 years, they *don't* mean that strontium loses all of its radioactivity after 56 years.

No matter what the original amount of strontium is—one gram or one ton—after 28 years, half of that original amount will remain. After another 28 years, one-quarter of the radioactive strontium will remain. After a third 28-year period, one-eighth of the original amount will be left. This is the decay process. Throughout the process, some of the radioactive material changes in composition. It actually changes into other, lighter elements. Radioactive decay continues at a steady rate until the amount of radioactivity becomes undetectable.

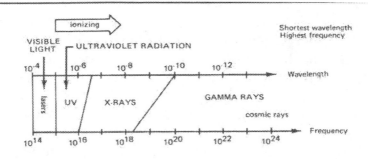

AM radio: approximately 535 to 1605 KHz (KHz = 10^3 Hz)
FM radio: approximately 88 to 108 MHz (MHz = 10^6 Hz)

Multiplying twenty times its half-life gives a substance's full *radioactive life.* For example, take strontium-90, with a half-life of 28 years. All but one part in a million will be converted into substances that are essentially stable, or not radioactive, in 560 years.

Terminology, here, is misleading. At first, I thought 28 years wasn't so long, until I understood what the "other half" of the half-life is. The radioactive life of plutonium-239 (half-life: 24,400 years) is nearly 500,000 years.

Types of Radioactivity

There are several types of radioactive emissions. Different radioactive elements can be characterized by the type or types of energy they release. *Alpha* radiation consists of particles that are identical to the nucleus of the helium atom (two protons and two neutrons), traveling at very high velocity. Alpha particles have a double positive (+ 2) electric charge. Alpha radiation is not very penetrating—it can be blocked by a few inches of air, a few sheets of paper, ordinary clothing, or unbroken skin—but if it gets into our bodies somehow, it will slowly irradiate the surrounding tissues, causing extensive damage over a period of time. The symbol for alpha emissions is the Greek letter α.

Beta particles move at an even greater speed than alpha. While they are released from the atomic nucleus, beta particles have the same mass and charge (-1) as electrons. Beta radiation can be blocked by a thin sheet of a metal such as aluminum or by a half an inch of wood. It can penetrate several millimeters of skin if an individual is not wearing heavy clothing. The symbol for this form of radiation is the Greek letter β.

X-rays and *gamma rays (γ)* are high-energy waves. They have short wavelengths, high frequencies, and great penetrating power—as anyone who has seen an x-ray picture knows. Barriers made of lead or concrete are necessary to shield the human body from unwarranted exposure to these forms of radiation. Most x-rays are produced artificially for use in medicine and industry, but some occur during the decay of unstable atoms in nature. They come from high-energy electrons. Gamma rays emanate from the atomic nucleus and can also either occur naturally or be produced artifically. X-rays and gamma rays have no charge.

While they don't have any special name, streams of neutrons, moving at high velocity, can also constitute a radiation hazard. This form of radiation, which lacks an electrical charge; is a product of nuclear power and technology. Nuclear physicists utilize neutron bombardment to initiate chain reactions. Neutron emissions have the most penetrating power of all. Several feet of water or concrete are required to block them.

The cosmic rays that reach our planet are composed of a mixture of high-energy gamma waves and atomic particles such as protons, neutrons and electrons.

How Is Radiation Measured?

Over half a dozen different units are commonly used to measure radiation. If you have problems when it comes to inches versus centimeters, you may very well feel tempted to skip over this topic! But by making the acquaintance of the various units, you can gain an enhanced understanding of what radiation is and get a clearer sense of the impact that radiation does have. Even if you're convinced that you'll never be able to distinguish between roentgens, rems and rads, just take a minute to see what's what.

Roentgen *(abbreviation: R)*. Named after the man who discovered x-rays, Wilhelm Roentgen (pronounced "rent-kin"), this unit was first used in reference to x-ray or gamma radiation exposure from x-ray machines. The usefulness of roentgens is limited, however, because they measure radiation's electrical charge in air. Rems and rads (discussed below) were developed to measure radiation energy in different situations.

Rem. Rem is short for *roentgen equivalent man.* It measures the degree of biological damage that ionizing radiation (alpha, beta, gamma, *or* x-ray) causes. Rems take into consideration the relative biological effectiveness of the energy that is absorbed by living tissue. One rem is roughly equivalent to one roentgen (1 R = .88 rem), and has the same biological effects.

Rad. Rad is short for *radiation absorbed dose,* and measures the radiation energy actually absorbed by the body. There are many units of measure for energy, including BTUs, calories, ergs, joules and watt-seconds. Historically, the *erg* is the unit that has been used to describe the energy delivered by ionizing radiation. The rad represents 100 ergs delivered to 1 gram of tissue. For beta, gamma and x-radiation, a rad is approximately equal to one rem. For alpha radiation, a rad is from ten to twenty rems.

RBE. The RBE, or *relative biological effectiveness,* compares the different health effects caused by different types of ionizing radiation. An alpha particle, for example, can have from ten to twenty times the RBE of a beta particle. This factor will vary, of course, depending upon specific circumstances, such as whether the exposure is internal or external.

LD. The toxicity of radiation can be expressed as a *lethal dose.* For instance, LD_{50} means a dose that is lethal for 50 percent of the exposed population. LD 30 50 means that 50 percent of the population will die within 30 days as a result of exposure. For humans, the LD_{50} is in the range of 400 to 500 rems. This calculation of lethal dose assumes that the population consists entirely of uniformly healthy adult males. In fact, however, populations affected by radiation are comprised of both sexes and mixed age groups, including infants and children and the elderly. There is a wide variation in health status among individuals as well. Thus, the LD—the amount of radiation that will kill half of the people within a given time period—can be much lower.

Amounts of ionizing radiation that are smaller than roentgens, rads, or rems are commonly referred to with the prefix *milli-,* which signifies "one thousandth." For example, a millirem (abbreviation: mrem) is one thousand times smaller than a rem. Roentgens, rads and rems measure radiation *exposure.* When there is a concern about the *rate* of delivery of radiation, these doses can be expressed per minute, per hour, or per year: for example, rads per hour, mrem/yr.

The *curie* (Ci) is the unit used to directly measure the radioactivity—quite literally, the radioactive activity—of a specified amount of a substance. The curie, which is named after Marie and Pierre Curie, the discoverers of radium, measures the activity of a source by counting the number of radioactive events that take place in a period of time. One curie equals 37 billion nuclear emissions (disintegrations) per second. By measuring and comparing curies, we can tell, for example, which of two substances is more radioactive. One gram of radium-226 has an activity of 1 curie. One gram of promethium-145 is many times more radioactive, as it has an activity of 940 Ci.

In addition to "milli-," other prefixes are sometimes used with the above-mentioned units (especially curies) to describe smaller radiation measurements: *micro*—(one millionth), *nano*-(one billionth) and *pico*—(one trillionth). A picocurie represents about two disintegrations per minute. You may recognize some of these prefixes from the metric system. As in the metric system, larger quantities may be expressed with the prefixes *kilo*-(one thousand) and *mega*—(one million).

As if all this is not confusing enough, an entirely different system of radiation measurements has been agreed upon by the international scientific community. In the years to come, we can expect to read about radiation measurements expressed in terms of *grays* and *becquerels.*

The gray (Gy) equals 100 rads, and will someday replace rads as a unit of measure. The Becquerel (Bq) is named after the French physicist Antoine Henri Becquerel, who discovered radioactivity. Equaling one radioactive disintegration per second, the Becquerel is a much smaller unit than the curie. It has been used in Europe as a system of measurement for about ten years.

The *sievert* (Sv) is a new international standard equaling 100 rems (100,000 millirems). For the present time, however, rems, rads, and curies remain the most commonly used terms in this country, and they are the terms I have used throughout this book.

Natural radiation has been estimated to contribute a dose of about 80 to 200 millirems per year to individuals in the United States. In addition, we average about 55 millirems per person per year from medical x-rays. Based on these two amounts, the National Council on Radiation Protection has established guidelines for radiation exposure saying that a person should receive no more than 5 rems in thirty years, excluding the amount received from background radiation and x-rays. Five rems over thirty years equals about 170 millirems (0.17 rem) per year. They suggest the highest amount in any one year should be 500 millirems (0.5 rem). Nevertheless, an increasing amount of evidence points to the fact that *there is no safe level* of radiation.

RADIATION IN THE ENVIRONMENT

When I told a friend I was writing a book about how to protect yourself from the harmful effects of the amount of radiation in the environment today, he answered: "Don't you understand man has lived and even evolved through radiation. How can you say it is bad? I've read doses we receive now from plant emissions are no more than the amount in the natural background radiation, so what are you getting all paranoid about?" I attempted to explain, but recognized I could have reacted the same way if I hadn't done the reading on the subject I have done in the last few years. Basically, it amounts to this: the rights of my friend—and of all the world's population now and for thousands of years into the future—have been trampled on by bureaucrats who have suppressed and distorted research on the fundamental dangers of radiation.

There are many sources of radiation. First of all, the earth has always been exposed to what is known as natural, or background, radiation. This includes the ultraviolet rays from sunlight, cosmic rays from outer space and also the radioactive decay of some of the earth's minerals. Altitude and geographic location affect the amount of background radiation we receive. In the United States, background radiation averages from 80 to 200 millirems annually.

Man-made radiation sources are proliferating; medical and dental x-rays are major culprits. Nuclear medicine is developing, along with such treatments as diathermy and radiation therapy. Air travel, video display terminals, tobacco smoke, building materials and television sets are often-ignored "everyday" sources of various types of environmental radiation. It is vital to consider these radiation sources along with the radioactive substances that are released into our environment by nuclear weapons tests and nuclear power plants. The effects of low-level radiation exposure on human health are *cumulative.*

X-Rays and Nuclear Medicine

It was not made clear until recently that despite their diagnostic value, medical and dental x-rays have harmful and cumulative effects on the human body. Now most people are aware that there is some risk to having dental x-rays, chest x-rays, CT scans or mammograms. These risks include tissue damage, sterility, and the development of cancer or leukemia. In addition, x-ray exposure during pregnancy can cause miscarriage or malformations in the infant. The variables that determine the risk include the part or parts of the body exposed, the patient's health status at the time, and the dose emitted by the equipment at the x-ray facility.

Since the 1940s, the field of *nuclear medicine* has rapidly developed the use of *radiopharmaceuticals.* Most of these radioactive materials are employed in making diagnoses. For instance, iodine-131 is used to evaluate the functioning level of the thyroid gland because of its characteristic of gravitating to the thyroid, where it can be observed to outline the gland's activities. Other common scans include the following: brain (technetium-99m), liver (gold-198), lung and blood volume (iodine-131), skeletal (strontium-85), vitamin B_{12} absorption (cobalt-60), spleen (mercury-197) and kidney (mercury-203). The half-lives of the radionuclides employed are between 6 hours and 65 days. While iodine-131 has a half-life of 8 days, its full radioactive life is 160 days—nearly half a year. Thus, its effects are longer-lasting than the half-life might lead you to believe.

Iodine-131 is also used in *radiation therapy.* John Gofman, a medical doctor and nuclear chemist whose pioneering research paved the way for the discovery of plutonium in the 1940s, estimates that individuals who have this treatment for thyroid malfunction or cancer will have about a 25 percent increased chance of developing cancer in the future. He also points out that the thyroid is not the lone recipient of the radionuclide. The whole body is irradiated via the bloodstream.

Perhaps the most well-known radionuclide is cobalt-60, which is sometimes used in the treatment of cancer patients. Other radionuclides may be injected directly into tumors (for example, iodine-125, iridium-192 and radium-226).

You can get comprehensive information on the risks associated with medical x-rays in the book *X-Rays: Health Effects of Common Exams,* by John Gofman and Egan O'Connor (Sierra Club Books, 1985).

Are You Exposed to Radiation in Your Job?

Many occupations cause extra exposure to radiation. Seven million workers in America are potentially exposed to ionizing radiation—including power plant employees, biologists, dental assistants and dentists, drug manufacturers, electron micro-scopists, fire alarm makers, food preservers, military personnel, nurses, pathologists, physicists, medical radiographers, pipeline overflow testers, radar tube makers, TV tube makers and uranium miners and processers.

Between seven and ten million people work in places where they are exposed to both ionizing and non-ionizing radiation from video display terminals (VDTs), the TV-like screens used in working with computers. Documented health problems among these workers have ranged from dizziness and fatigue to severe headaches, nausea and cataracts. An unusual number of miscarriages and babies born with birth defects have been reported by women working with VDTs. So far, however, no one has been able to test whether these problems were radiation-induced.

There is a newsletter called *VDT News,* which addresses itself solely to the health and safety concerns of VDT users. The January/ February 1986 issue reported a newly discovered effect: the tendency of red blood cells to clump together after a person spends just five minutes in front of a VDT screen. This phenomenon, known as Rouleau clumping, causes a restriction of blood flow and limits red cell efficiency. A study published by Dr. John Ott, in the *International Journal for Biosocial Research* (July 1985) reported that a blood sample from a VDT user showed Rouleau clumping after a full workday, but after the employee spent the next day outdoors, the blood sample showed no clumping.

The authors of a booklet entitled "Don't Sit Too Close to the TV: VDTs/CRTs and Radiation" offer the following perspective: "We see VDT radiation as part of the increasingly hazardous problem of 'electronic smog,' in which the radiation by-products of electronic technology pollute the work-place and the environment with potentially devastating health and genetic effects."

Nevertheless, the possibly serious hazards of both ionizing and non-ionizing radiation in the workplace have repeatedly been downplayed. According to a December 1983 article in the *Village Voice,* the National Institute for Occupational Safety and Health (NIOSH) issued a report in 1975 stating that VDT x-ray emissions and background radiation levels were equal. However, "the background levels they measured exceeded by over 40 per cent the highest levels ever recorded for natural radioactivity." The article also questions the recommendations NIOSH made in July 1983. While calling for improvements in the design of VDT workstations to reduce eyestrain, "The NIOSH recommendations state 'the VDT does not present a radiation hazard to the employees working at or near a terminal.'"

Microwave and low-frequency radiation is all around us—emitted by television and radio towers, satellites, high-voltage electric lines, police radios, paging systems, microwave ovens and electronic games (see below for a more complete list). It is not ionizing—it doesn't displace electrons and start biochemical reactions in our cells—but it does affect human health. Reports about the damaging effects of microwave radiation come from many medical specialties, including behavior and physiology, immunology, neurology, ophthalmology, genetics, endocrinology, hematology and the study of the cardiovascular system.

Paul Brodeur, the author of a book on the effect of radiofrequency energy, quotes a 1971 government report that warned: "The consequences of undervaluing or misjudging the biological effects of long-term, low-level exposure could become a critical problem for the public health, especially if genetic effects are involved.'" He argues that "in spite of the denials and claims to the contrary that flow from the military-electronics industry complex, the microwave and radio-frequency radiation problem is not a limited threat. Indeed, the microwave radiation problem affects virtually every man, woman and child in the land."

Brodeur's book, appropriately titled *The Zapping of America,* sounded the alert in 1977. In 1984 the Environmental Protection Agency announced that it was considering a plan to limit the power of radio and television transmitters. The EPA was quoted in the journal *Science:* "Recent animal studies have shown that weak electro-magnetic fields can produce subtle changes in the nervous and immune systems, in blood, and in behavior."

Microwave Radiation

Sources

- alarm systems
- CB radios, hand-held walkie-talkies, and other radio communications systems (for example, those used in large buildings)
- cellular telephones
- diathermy units and other medical instruments
- electronic games
- remote control garage doors
- microwave ovens
- paging systems
- radar devices (for example, those used in boats and cars)
- satellite dishes
- signal generators

Professions with Particular Exposure Risks

- communications workers
- electrical engineers
- electricians
- medical technicians
- telephone repairers
- utility workers

Health Damages

- birth defects and stillbirths ("problem pregnancy clusters")
- cataracts
- changes in blood-brain barrier and calcium flow
- dizziness
- disturbances in function of endocrine glands
- fatigue
- genetic damage
- headache
- heart damage
- impaired judgment increased lymphocyte count
- irritability
- leukemia
- muscular weakness
- skin burns

Is Your House Radioactive?

Uranium naturally found in the earth may be a component of building materials such as cement or stone; if these are used to build your house, you will be subjected to radon, the radioactive gas that is emitted from uranium. To a certain extent, exposure to this natural substance is unavoidable. Unfortunately, improved weather sealing, designed to conserve energy, reduces the circulation of air and intensifies the potential effects of the radon.

Aside from building materials, radon may come from underground geologic formations. Chief among these formations is uranium-rich granite rock, which comprises the Reading Prong underlying parts of Pennsylvania, New York and New Jersey. Many other areas of North America also stand on granite bedrock. Radon can seep through cracks in basement floors and walls or be brought in through water pipes. Rising columns of heated air can also draw radon into a building. Fortunately, however, radon levels can be measured and steps can be taken to seal cracks and vent the radon away from the building.

The radon problem is much more widespread than was once thought; according to a November 15, 1985 article in *The New York Times,* Federal health officials have identified radon as the second leading cause of lung cancer in the United States. According to the *Times,* unsafe levels of radon may endanger from one million to eight million homes across the nation. The Environmental Protection Agency says that as many as 12 percent of all homes in the United States may contain cancer-causing levels of the gas. In August 1986 the EPA suggested a radon exposure limit of 4 picocuries per liter of air.

Cigarette Smokers Inhale Radiation

Information known in scientific circles since 1964 (when polonium-210 was discovered in tobacco by Vilma Hunt, a researcher at Harvard University) is surfacing to the public in the mid-80s. The fact is that cigarettes contain astonishing amounts of radiation and that a person who smokes about a pack and a half a day annually receives an ionizing radiation dose to lung tissue that's the equivalent of about 300 chest x-rays per year!

The source of this radiation is a fertilizer which is used on the tobacco plant and contains phosphates high in uranium. This uranium decays to radium and then to polonium-210. This substance is absorbed by the tobacco leaves, goes into the cigarettes, and then passes into your lungs, where it can disrupt cells and cause damage to the immune system.

Although polonium is known to be a carcinogenic (cancer-causing) agent, it is difficult to estimate the number of cases of lung cancer that it could cause. One reason cigarette smokers are more susceptible to radiation damage is that the tobacco tars immobilize the cilia (the tiny hairs lining the windpipe and the inside of the lungs). In healthy individuals, the sweeping motion of the cilia helps expel contaminants.

But as a result of continual smoking of cigarettes, pipes or cigars, the radioactive particles, tars and many powerful chemicals accumulate and overwhelm the body's attempt to cleanse the lungs.

Instead of being eliminated, these substances are carried by the blood to other parts of the body—the liver, kidneys and lymph nodes. In addition, they may settle in plaques lining the arteries, contributing to atherosclerosis and strokes.

It has recently been recognized that nonsmokers who are near smokers also absorb some percent of the radioactive poisons when they inhale.

There is also some evidence that polonium-210 reacts *synergistically* with other carcinogens, for example, benzopyrene and radon's decay products (which are particles known as "radon daughters"). This means that the effects of each substance are intensified. Benzopyrene is present in cigarette smoke, and radon (as we have seen) is an environmental hazard in many buildings. Radon daughters are attracted to smoke particles, and in their presence, remain airborne rather than settling on walls and floors, affecting smokers and nonsmokers alike.

As if all this wasn't enough, cigarette smoke has also been found to contain radium-226, lead-210 and potassium-40.

Smoke detectors, which in some urban areas are required by law in all apartment dwellings, can be another source of radiation in the home. There are two types of smoke detectors: ionization and photoelectric. Ionization smoke detectors contain from 1 to 5 microcuries of americium-241, a radioactive element that is a product of power plants and military reactors. Americium-241 emits alpha particles and gamma rays continually. In contrast, the photoelectric smoke detector emits no radiation. While many fire experts agree it is better suited to smoke detection, it is about twice as expensive as the ionization type.

The use of ionization smoke detectors poses several dangers, including leaks from faulty detectors and environmental contamination after disposal. Ionization smoke detectors have a practical life of about five years. After their disposal, the americium might pass into the water supply, go into the food chain or be released into the air if burned with garbage. About 40 million ionization smoke detectors have been installed around the United States. A conservative estimate is that there is 1 microcurie of americium-241 per detector (some contain up to 4 or 5 microcuries). If this amount were to be evenly distributed and inhaled, according to radiation chemist Dr. Chauncey Kepford, it would have the potential to cause about 300 cases of lung cancer. Smoke detectors may contain small amounts of americium, but they add to our insidious, ubiquitous, constant and cumulative exposure to health-limiting radiation.

The Atomic Age

In 1905, Albert Einstein, a gentle man who liked to play the violin, started us on our present path when he formulated the theory of relativity—that matter could be converted into energy. Fascinated by the new territory opened up by this theory, scientists such as James Chadwick (who discovered the neutron), Niels Bohr, Ernest Rutherford, Enrico Fermi and Leo Szilard dove into further explorations of the atom.

They discovered that certain atoms could be split, turning a tiny quantity of matter into a great quantity of energy. This process is known as *fission.* When uranium is bombarded with a stream of neutrons moving at a certain speed, it splits, releasing energy and starting a chain reaction. During World War II, President Franklin D. Roosevelt, key military leaders and scientists became convinced that in order to win the war the U.S. needed to be the first nation to develop a weapon utilizing the atom's explosive power.

The first atomic chain reaction was initiated in 1942, but the Atomic Age truly began on July 16, 1945 when the first test bomb was exploded at Alamogordo in New Mexico. This was followed in early August by the two devastating explosions in Japan. Despite the immediate opposition of some of the very people who had helped to develop the bomb, there was no retreat from the Atomic Age.

Since that time, several other nations (including the U.S.S.R., Great Britain, France, China and India) have developed nuclear weapons. Over twenty-five countries have begun construction or operation of nuclear power plants. Approximately 375 commercial reactors around the world were in operation as of mid-1986, 101 of these in the United States.

As the April 1986 accident at Chernobyl Unit 4 in the Soviet Union made clear, the truth is that nuclear fallout from power plants and bomb tests does not respect national borders or political differences. Because of the military applications of nuclear technology, however, there is a scarcity of information regarding the world's nuclear resources. Throughout this book, I have focused on the United States, because the most reliable information readily available to the general public deals with this country.

Nuclear Power/Nuclear Weapons

In nuclear reactors, a moderator, which is usually water, is added to keep the chain reaction from going too fast. Heat buildup is controlled by certain materials, such as cadmium, which soak up extra neutrons that could cause too many reactions. Rods of cadmium are inserted into the fuel to remove as many neutrons as are needed to obtain specified amounts of heat, which is converted into electricity.

Plutonium, one byproduct of uranium fission, is also readily split. One of the most toxic substances known, it is now put to use as the "trigger" for atomic bombs, a power source for space rockets and satellites, and in nuclear reactors.

In the 1981 book *Radiation and Human Health,* Gofman notes:

> Virtually all the plutonium injected into the atmosphere [from bomb testing] through 1962 has now returned to the earth. The best estimate of the total global dispersion and fallout of . . . plutonium is 320,000 curies. . . . As all the plutonium from the nuclear testing descended through the atmosphere, attached to a variety of particulates, it was available for inhalation by human beings, particularly in the Northern Hemisphere. This is the reason that virtually every human being on earth now has plutonium in his or her body.

In 1975, about 6 tons of plutonium were produced as byproducts of nuclear power plants. The total stockpile of plutonium is actually much greater. In July 1984, *Science Digest* printed an article by James Chiles, stating:

> An inventory of U.S. plutonium would probably reveal a total of about 190 tons: perhaps 100 tons in reactor fuel

and spent-fuel rods, 80 to 90 tons in weapons, less than a ton in the environment and in our bodies and a few pounds in electrical power sources for satellites and heart pacemakers.

According to Gofman, *each* pound (if finely divided and inhaled) could cause up to *42 billion* cancer cases. As the world's population now numbers about 3 billion, this estimate also takes account of future generations. The plutonium we have liberated since 1942—from nuclear plants and nuclear weapons tests—will contaminate the environment for tens of thousands of years to come.

In addition to plutonium, many radioactive substances are byproducts of fission reactions. Some reach a stable state quickly and pose no threat; these are germanium, gallium and zinc. Fairly short-lived but potentially dangerous is iodine-131, which has a half-life of 8 days. Other isotopes are long-lived and a source of radiation exposure for a long time and over a long distance: among these are cesium-137, which has a half-life of 30 years; strontium-90, which has a half-life of 28 years, and plutonium, which has a half-life of 24,400 years. The chart on pages 20-21 summarizes some of the radionuclides commonly emitted by nuclear power plants into the air. Radioactive substances are also released into the plants' cooling water.

The development of nuclear weapons and nuclear power has gone hand-in-hand in the United States. In 1946 the U.S. Government released the first official report on nuclear development, and proposed that complete control be given to the new Atomic Development Authority (the first of what has proved to be a baffling series of commissions) to set out on an ambitious weapons and energy program.

A 1952 Atomic Energy Commission (AEC) document, recently declassified, discussed the feasibility of utilizing nuclear reactors for a dual purpose: to produce plutonium for the American military and at the same time utilize the heat generated in the process to make electricity for commercial use. By bringing in private industry, the government planned to reduce costs. Industry saw profits coming from the sale of plutonium rather than electricity. Economic benefits appealed to utilities and government; the AEC boasted that it had found the perfect energy source and that this power would be even too cheap to meter.

President Dwight D. Eisenhower made a speech at the United Nations in 1953, with a proposal for "Atoms for Peace." The stated purpose of this plan was to promote peaceful uses of atomic energy "for the benefit of all mankind." Eisenhower later acknowledged that the plan was also designed to allay the public's fears and reservations about the bomb and its destructive potential. Thus began a trail of deception.

The first commercial nuclear reactor in the United States was started up in 1958. President Eisenhower presided over a ceremony in the Oval Office of the White House to commemorate the occasion, just five years after he had proposed the peaceful

Half-Life of Selected Radionuclides Emitted by Nuclear Facilities Into the Air

Substance	Half-Life
Phosphorus-32	14.3 days
Phosphorus-33	25.0 days
Chromium-51	27.8 days
Manganese-54	303.0 days
Iron-55	2.6 years
Iron-59	45.1 days
Cobalt-58	71.3 days
Cobalt-60	5.3 years
Nickel-63	92.0 years
Niobium-92m	10.1 days
Tin-117m	14.0 days
Tungsten-185	75.8 days
Uranium-237	6.8 days
Krypton-85	10.8 years
Xenon 129m	8.0 days
Xenon 131m	11.8 days
Xenon 133	5.3 days
Iodine-129	1.7×10^7 years

Iodine-131	8.1 days
Strontium-89	52.0 days
Strontium-90	28.1 years
Yttrium-91	58.8 days
Zirconium-95	65.0 days
Niobium-95	35.2 days
Ruthenium-103	39.6 days
Ruthenium-106	367.0 days
Tellurium-127m	109.0 days
Tellurium-129m	34.0 days
Cesium-134	2.1 years
Cesium-136	13.0 days
Cesium-137	30.2 years
Barium-140	12.8 days
Cerium-141	33.0 days
Cerium-144	284.9 days
Neodymium-147	11.1 days
Plutonium-239	24,400.0 years
Plutonium-240	6,580.0 years
Americium-241	458.0 years
Curium-242	163.0 days
Curium-244	17.6 years

Note: The "m" following the atomic number of a radionuclide designates a *metastable* substance. Other forms of the radionuclide have different half-lives.

Sources:
Selection of radionuclides from *Radiation Exposure to the Public from Radioactive Emissions of Nuclear Power Stations* by B. Franke, E. Kruger, and B. Steinhilber-Schwab of the Institute for Energy and Environmental Research, Heidelberg, West Germany. Used by permission.
Values for half-lives from *Handbook of Chemistry and Physics,* 63rd edition. Weast, Robert C, Ed. Boca Raton, FL: CRC Press, 1982.

use of atomic energy. Nevertheless, reactors had been operating since the 1940s to produce plutonium for nuclear weapons, which were being tested at a rapid pace throughout this period of time.

Nuclear weapons manufacture and testing have added an enormous amount of radiation to the global atmosphere. The residue of 817 U.S. nuclear tests—both atmospheric and underground—from 1945 to 1985, remains. (The global total during this period of time was 1,625 tests.) In the United States, underground tests continued at the rate of about one a month through spring 1986. As authors Harvey Wasserman and Norman Solomon have pointed out in their landmark book *Killing Our Own:*

> One of the most pervasive—and erroneous—beliefs about the U.S. nuclear testing program is that its radioactive fallout ended when the Limited Test Ban treaty took effect in 1963. When the nuclear tests went underground, people assumed the weapons testing radiation threat disappeared. This comforting notion, carefully nurtured by the government, is false.

In 1953, ranchers in Utah lost 4,500 sheep after bomb tests were conducted nearby (at what is now known as the Nevada Test Site, or NTS). They sued for reparation in 1956, but lost the case. This decision was reversed by the courageous decision of Federal Judge A. Sherman Christensen, who had presided over the original trial. In 1982, he ruled that the government of the United States practiced fraud upon the court in the 1956 trial. Christensen said that government agents made false statements, that witnesses were pressured, that crucial information pertaining to the case was withheld and that court proceedings were manipulated to the government's advantage. In an article in *The New York Times* (August 15, 1982), a rancher explained why this was so:

> I'm sure they felt if this whole can of worms was opened up beginning with sheep, the people would rise up against the testing.

Christensen's order for a new trial has since been overturned, and a series of appeals has followed. As of this writing, the ranchers' case remains unresolved. Although aboveground testing at the NTS has ceased, the residents of the area, who call themselves "downwinders," still suffer from radiation-induced cancers. According to professor Howard Ball, author of *Justice Downwind*, as recently as 1974—sixteen years after the last aboveground blast in the area—the incidence of childhood leukemia downwind of the Nevada Test Site was one-and-a-half times the national average.

Despite repeated indications that nuclear technology is fraught with hazards to human health and the environment, there are plans for setting up nuclear power plants in outer space. These plants would provide fuel for space stations, planetary bases, power for deep space voyages and laser weapons. The Department of Energy and the Defense Advanced Research Projects Agency agreed in 1983 to together manage a space-reactor development program aiming to have five to ten power stations in orbit by the year 2000. The funding of $13.6 million in the year 1984 for the reactor project was then incorporated into President Ronald Reagan's Strategic Defense Initiative (the "Star Wars" program). In 1983, the amount of $6.4 billion was allocated to military space projects. A full-scale space-based missile defense system would require hundreds, perhaps thousands, of large nuclear reactors in continuous permanent orbit around the earth.

Hans Bethe, winner of the 1967 Nobel Prize for physics and a widely respected researcher, now works to stop the development of nuclear weapons and is against space-based missile defense. In an interview in 1984, he said: "People want to eliminate the danger of nuclear weapons by technical means. I think this is futile."

Large-Scale Problems

Nuclear power production is an insidious and increasingly all-pervading source of radiation that will affect our present population and our descendants for thousands of years. In many cases, facts that have been concealed from the U.S. public include basic design deficiencies and the use of inferior materials. The construction and operation of these plants is rife with carelessness and negligence and the possibility of human error at every level. There are planned daily emissions of "safe" amounts and unplanned-for leaks from all the plants.

In the 1970s, many leaks went unreported. During the time I was researching this book, the few accounts I came across were tucked in the middle of *The New York Times.* One *Times* article ("Nuclear Leak Data Criticized," p. A17, April 17, 1984), I quote in its entirety:

> Poor reporting of a leak at Florida Power and Light's Turkey Point nuclear plant has led to a nationwide warning from the Nuclear Regulatory Commission. On Jan. 5, technicians at the plant told the commission about an accidental release of radioactive gases that lasted 15 to 20 minutes. But commission officials said the utility submitted the wrong figures. They said, based on the figures, that the release was potentially dangerous and should have put the plant on emergency alert, prompting evacuations within two miles.

According to another article in the *Times* ("Carolina Nuclear Plant in Big Radioactive Leak," Sept. 4, 1984), a plant spokesman admitted that "the release was estimated by plant officials at 50,000 curies. Company officials say that any release of radioactive material over 10,000 curies is considered a 'significant release' and that the plant has had five releases of that size in 10 years."

The much-disputed Diablo Canyon plant in California, built close to an earthquake fault, sprung a leak in one of its reactors on the opening day, April 23, 1984; the opening was postponed.

Plutonium escapes from the Rocky Flats nuclear weapons plant in Colorado during everyday operations, finding its way through the limited filtering system in the smokestacks. The Department of Energy acknowledges that there is a certain level of plutonium in the surrounding soil. About 600,000 people live within twenty miles of Rocky Flats, and the cancer rate is higher in that vicinity than in the rest of the state. The plant workers have a cancer rate five times that found in the rest of Colorado.

A note here: Since 1953, when Rocky Flats opened, they've made 26,000 plutonium triggers for bombs, at the rate of three per day. Four pounds of plutonium are used in each trigger. If that four pounds were to be evenly distributed, divided into fine particles and placed in the lungs of people among all the world's population, it would have the potential to kill everyone on the planet.

Documents made public in 1986 revealed that intentional releases of radioactive iodine from the government-owned Han-ford nuclear reservation in the 1940s were of a magnitude that today would be considered a major nuclear accident. In addition, these releases were followed by a series of plutonium reprocessing accidents in the 1950s and 1960s.

America's nuclear plants have had critical accidents; often, luck has played a role in keeping them from escalating into something worse. A few examples: two major fires at Rocky Flats released enough plutonium to build seventy-seven bombs. In 1961, an explosion at a reactor in Idaho killed three people and exposed over 700 to high levels of radiation. A 1975 fire at the Browns Ferry nuclear power plant in Alabama raged out of control for hours before it could be extinguished. In February 1983 there were two incidents at a Salem plant in New Jersey. The circuit breakers meant to shut down the plant failed to work automatically twice within one week. The plant was on low power then, but if it had been going at full power, the Nuclear Regulatory Commission said, "an extremely severe accident" could have resulted.

Although the Three Mile Island fiasco in 1979 (which riveted my attention and led me to begin researching this book) was not referred to as a major accident, we now know that it went to within 60 minutes of a complete meltdown.

A plant inspector employed at Diablo Canyon, Charles Stokes, was dismissed after he voiced his concerns about safety. In an interview in *The New York Times* (April 30, 1984, page 1), he predicted: "It may take 20 years but I feel it's a certainty that we'll have a major leak of radiation or a meltdown."

These sobering close calls have pointed out the fact of basic flaws in plant design and materials and also the potential for human error. There was an average of 6.5 sudden violent plant shutdowns (known in the industry as "scrams") per plant in 1984 and 1985. Although the average remained constant, the *number* of scrams increased—from 518 in 1984 to 601 in 1985. Each represents potential trouble. In April 1985, the Nuclear Regulatory Commission estimated the chance of a severe core meltdown at a reactor in the U.S. before 2005 at about 45 percent *(The New York Times,* April 17, 1986). The chance of an accident destroying a reactor anywhere in the *world* within the next ten years has been estimated at 86 percent *(The New York Times,* October 27, 1986).

Sloppy construction, poor-quality materials criticized as substandard, weak welds in the reactor, and cracks in the pipes are faults found in most plants. On January 23, 1983, a front-page story in *The New York Times* reported that the NRC had learned there were "hundreds and perhaps thousands of substandard and fraudulently marked small steel components. In one case, the commission has determined pipe fittings, sold as being able to withstand 3,000 pounds of pressure, were expected to endure only 150 pounds."

In 1978, the Union of Concerned Scientists managed to get some files from the NRC with data about plant safety weaknesses. The documents, referred to collectively as "The Nugget File," listed many incidents of equipment failure and personnel failure that had been noted by the NRC and kept secret:

> For want of fuses, key nuclear safety equipment is rendered inoperative. Electrical relays fail because they are painted over or welded together or disconnected . . . . Simple maintenance operations disrupt established safety precautions, as when valves that affect the water level in the reactor were accidentally shut off during the repair of a leaky faucet in a plant laboratory. Sensitive pieces of safety equipment malfunction because they are frozen or burned or flooded or dirty or corroded or bumped or dropped or over pressured or unhinged or miscalibrated or mis-wired; they are also shown to cite one of Hanauer's marginal comments, "to be guaranteed not to work" because of initial bad designs.

A major issue that has emerged is *containment:* the effectiveness of the structures designed to prevent the escape of radiation from the reactor building. Robert Pollard, a nuclear safety engineer for the Union of Concerned Scientists, was quoted in a June 1986 press release by Public Citizen—Critical Mass Energy Project as stating that 39 nuclear plants in the U.S. have pressure-suspension containment systems similar to the one that failed at Chernobyl in the U.S.S.R. "The basic problem," he explained, "is that their containment building might rupture—or fall down—under the stress of a major accident." Major accidents are not limited to meltdowns. Other possible causes of disaster include steam explosions, runaway reactions and rapid fuel breakup.

To compound all this, personnel are often undertrained or inadequate for such responsibility. For instance, as one news story from San Onofre, California revealed: "Thirteen guards hired to protect the San Onofre nuclear plant have been dismissed for having 'significantly high levels' of drugs in their bodies, officials said."

A *New York Times* article with the headline "5 to 6 Operators Fail to Pass Test at Largest Atomic Plant" (January 28, 1984) tells that a surprise Federal examination given to six licensed nuclear reactor operators was failed by five (who were then removed from duty). The examiners explained, "Like the pilot of an airplane, the operator is in charge of the reactor, monitors the controls and decides what to do in case of malfunction or other emergency is indicated."

The magnitude of the problem becomes even more apparent when we consider that incompetence, human error, and design flaws are by no means limited to the United States. Standards for design of nuclear plants, safety precautions, and training of personnel vary in different parts of the world. Portions of a September 1985 report prepared by the U.S. General Accounting Office were declassified in the wake of the April 1986 Chernobyl disaster. *The New York Times* reported that the agency had a record of 151 "significant nuclear safety incidents" between 1971 and 1984 in fourteen Western countries.

Nuclear Plant Leaks Endanger Everyone

In the 1960s the Atomic Energy Commission authorized a study by two eminent scientists: Dr. John Gofman and Dr. Arthur Tamplin. Their task was to investigate the health effects of peaceful applications of radiation. They presented their findings in October 1969 at a symposium in San Francisco, stating that low-level radiation is much more serious than had been thought. They urged a tenfold reduction in the maximum permissible radiation dose for the general public. This information was not well accepted, and Gofman and Tamplin soon lost their government funding for research of any kind.

In 1971, they published their views independently, in the enlightening book *Poisoned Power,* in which they state:

> If the average exposure of the U.S. Population were to reach the allowable 0.17 rads per year average, there would, in time, be an excess of 32,000 cases of fatal cancer plus leukemia per year, and this would occur year after year.

Dr. Irwin Bross, a biostatistician at the Cancer Institute in Buffalo, New York, made clear his opinion on government regulation of allowable standards for radiation workers in the book *Radiation Standards and Public Health:*

> There was no science in setting the 5 rem level; there was no science in setting the [previously accepted] 50 rem level. These are arbitrary values as you have heard. They do not indicate safety.

> The effects of radiation on workers may have been concealed. A small article appeared in *The New York Times* on October 7, 1984, under the title "Radiation Tests Are Assailed":
> Four union leaders representing hundreds of workers in nuclear weapon plants have accused the Department of Energy of manipulating test results to minimize the health effects of radiation exposure in the plants, and have asked that responsibility for the tests be shifted to the Department of Health and Human Services . . . .
> They accused the Energy Department of the "manipulation of important health effects research"

that they say "indicate that serious health problems are occurring at several DOE nuclear facilities."

Prior to the publication of *Poisoned Power,* there had been little public consideration of the idea that nuclear plants could leak radiation into the environment. Gofman and Tamplin showed that this is, on the other hand, a fact and a threat to the population, especially to those living near a plant or eating food or milk that comes from near a plant. In their thorough account of the health danger posed by low-level radiation, the 1982 book *Killing Our Own,* authors Harvey Wasserman and Norman Solomon explain:

> Before 1969 only a tiny handful of scientists had considered the issue of leaking reactors at all. The leaks in general came from the breakdown of fuel sheathing in the controlled but superhot reactor core. As cooling water flows around the core, it picks up radioactive isotopes, itself becomes radioactive, and carries that through the maze of pipes and valves around the plant.
>
> Some of the emitters then escape through the plant stacks as gases and particulate matter, in particular lethal isotopes of iodine, strontium, cobalt, carbon, cesium, and noble gases. Some—particularly tritium—are also flushed out with waste water and into local rivers and the oceans. Some neutrons and gamma radiation also penetrate the containment vessel that tops the reactor. Such releases are an ongoing aspect of normal power reactor operations . . . .
>
> Both the scientific community and the public had been assured that reactor leakage would be virtually nonexistent, and at any rate would pose no serious health threat.

In a 1976 article in the *Journal of the American Medical Association,* entitled "The Plutonium Controversy," Gofman focused on one aspect of the health threat posed by nuclear power plants:

> Plutonium is a uniquely potent inhalation carcinogen, the potential induction of lung cancer dwarfing other possible toxic effects. . . . it is my opinion that plutonium's carcinogenicity has been seriously underestimated. If one couples the corrected carcinogenicity with the probable degree of containment of the plutonium, it appears that the commercialization of a plutonium based economy is not an acceptable option for society.

When we consider that, as Gofman states, "for Pu-239 the average half-life is equal to 24,400 to 34,200 years," we begin to recognize what we are up against.

Do You Live Within 30 Miles of a Nuclear Facility?

As of 1980, about 30 million Americans lived within thirty miles of a nuclear power or weapons plant. Because of plant emissions and leaks, people living within a thirty-mile radius receive a higher dose of radiation than those living further away.

The map on pages 30-31 shows some, but not all, of the commercial and military nuclear facilities in the United States. The military facilities shown here include nuclear material production facilities, nuclear weapons production and assembly plants, design laboratories and landfills. The commercial facilities include power plants, uranium mines and mills, and both intermediate and long-term waste storage areas. "Waste brokers" store low-level radioactive waste for several months to a year before shipping it to more permanent burial sites.

As you look at the map, keep in mind what it does *not* show. It does not show the relative sizes of the nuclear facilities or waste sites, nor the amount of radioactivity they involve. A few examples will give you some idea of the size of the problem: the Hanford Reservation in the state of Washington is 570 square miles (over half the size of Rhode Island); some uranium mill tailing sites in the West are as large as 300 acres; and a repository

Map of Commercial and Military Nuclear Facilities in the U.S.

Military Facilities {

△ Nuclear material production

◆ Nuclear weapons production

▲ Design laboratories

W. Waste landfills

T. Current bomb test site

★ Low-level waste broker facilities

◇ Waste facilities

By #. Uranium mills/mines: number of facilities in each state is given

Nuclear Power Plants, Storage and Waste Sites

●   Operating reactors (including operating high-level waste sites)

○   Reactors under construction

■   Operating low-level waste sites

□   Closed low-level waste sites

×   Away-from-reactor spent fuel storage sites

*   Sea disposal sites (pre-1970)

for radioactive waste that is being built near Carlsbad, New Mexico, covers 19,000 acres.

In addition, the map does not include the over 100 missile silos and nuclear weapons depositories that are scattered across the nation. These locations include Army, Navy and Air Force bases. While the precise details are classified, information on the deployment of nuclear weapons is available from the Center for Defense Information in Washington, DC. The War Resisters League in New York City has a comprehensive map of "Nuclear America," describing a variety of nuclear facilities and specifying their owners or contractors.

The map included in this book does not show the locations of nuclear reactors used for research, and it does not show the thousands of other research facilities that use radioactive materials. These include industrial and military laboratories, hospitals, medical centers, food processing plants, colleges and universities. Moreover, it does not indicate the location of radioactive material that has been secretly buried by companies that did not think or care about the consequences. If it *did* show all these things, perhaps the blackened map would indicate that there are no safe places to hide.

Dr. Carl J. Johnson, the Secretary of Health for South Dakota and a former county health commissioner in Colorado, has written extensively about the health effects in populations living near a nuclear installation. In a letter to the *American Journal of Public Health,* he took exception to the estimated amounts of radiation found nearby and also to the recommendations that resulted from these estimates:

> I urge that an attitude of rigorous and critical scientific inquiry be applied to each link of the chain of assumptions upon which risk estimates of radiation exposure are developed, and upon which nuclear agencies place reliance on the promulgation of their doctrine concerning radiation hazards.

The United States Department of Energy has three facilities at Oak Ridge, Tennessee. One makes nuclear components for weapons. A study by the Tennessee Valley Authority, reported in *The New York Times* on January 24, 1984, found more than 140 dangerous chemicals and radioactive substances in the creek sediment around the city. Two forms of plutonium were found.

In 1975, a large amount of strontium-90 was discovered in the milk at a farm near the Shipping port, Pennsylvania plant. In 1977, Dr. Ernest Sternglass found a correlation between the strontium emissions from the Millstone plant at Waterford, Connecticut with high cancer incidence nearby.

Waste

One of the legacies of the nuclear era that did not receive due attention from policy-makers throughout the first thirty years of intense nuclear power development is radioactive waste. The Federal Government spent many billions of dollars to develop this new energy source and produce nuclear weapons during the 1950s and 1960s, but during the same period, only $300 million was spent to research a safe way to store radioactive waste products.

Yet the accumulated radioactive waste from nuclear power plants and weapons production poses a threat for periods of time up to hundreds of thousands of years. Only in the last few years has the Federal Government attempted to come to grips with this pressing problem. We have not yet found a convincing technology for containment or long-term storage. Waste we have tried to store underground in the Western United States has dribbled into the soil and the water table. Up until 1970, large amounts of radioactive material were tossed into the Atlantic Ocean—in casks that will disintegrate in a number of years.

Radioactive wastes are classified according to their degree of radioactivity. The two general categories are low-level and high-level. The amount of contamination per gram of material at the time of packaging and transport is the factor the nuclear industry uses to classify wastes. (See *A "Low-Level" Nuclear Waste Primer,* © 1985 by the Sierra Club Radioactive Waste Campaign.)

"Low-level" is a misleading term because it minimizes the potency of the waste it is referring to, especially when large quantities of such waste accumulate. Low-level radioactive waste (LLRW) can range from slightly contaminated clothing (gloves, aprons, etc.) and equipment to highly radioactive sludge from nuclear reactors.

Actually, low-level radioactive waste is practically everything except the reactor's fuel assembly and the spent nuclear fuel (sometimes referred to as *irradiated* fuel) itself.

Two main low-level radionuclides are cobalt-60 and cesium-137, with half-lives of five and thirty years and radioactive lives of 100 and 600 years, respectively. During this time they must be safely contained. Other radionuclides, especially those that result from the decontamination of reactor pipes, have half-lives of thousands of years. They are also considered low-level waste and are thrown into landfills with minimal protection.

Nuclear power plants are not the only source of LLRW. The United States military is also a producer. In addition, hospitals, medical research facilities and a variety of industrial settings generate an enormous volume—estimated in 1983 to be somewhat over 80 million cubic feet. The Environmental Protection Agency has projected that by the year 2000 about one billion cubic feet of low-level waste will be generated. This volume could cover a four-lane highway from New York to California with a layer one foot thick! Yet, in addition to the massive volume of this LLRW, policy-makers in the years to come will need to consider the half-life and radioactive intensity of the various substances comprising the waste. The radionuclides used in medicine, for example, generally have half-lives measured in hours or days, and pose less of a storage problem than wastes from reactors.

The generation of waste begins at the beginning of the nuclear fuel cycle with the mining of uranium. In the U.S., the primary uranium-mining states are Wyoming, New Mexico, Colorado and Utah. Approximately 40 tons of uranium oxide are mined daily. This uranium is extracted from 20,000 tons of ore; thus, it can be seen that uranium processing leaves behind a huge amount of debris, which is a source of radium, thorium and radon. In 1983, about 175 million metric tons (a metric ton equals 1,000 kilograms, about 10 percent more than 2,000 pounds) were left at 27 sites, some as large as 300 acres, in the states where uranium was mined. Uranium *tailings* (as such debris is known) are sometimes classified as low-level waste but usually considered separately. The fact—learned from hard experience—is that abandoned uranium mills remain too radioactive to be suitable for development as residential housing sites. The

Uranium Mill Tailings Radiation Control Act of 1978 recognized this fact, and identified two dozen sites requiring cleanup.

"High-level" radioactive waste is composed of the spent fuel assemblies removed from nuclear reactors and also the waste that results from reprocessing of spent reactor fuel. Defense reactors always reprocess spent fuel but commercial power plants stopped doing this in the early 1970s. High-level wastes contain deadly concentrations of some of the most formidable poisons existing, including plutonium, cesium-137 and stron-tium-90. Spent fuel assemblies, which are the heart of nuclear reactors, may add up to about 100 metric tons of high-level radioactive waste by the end of the century.

Congress has passed two important pieces of legislation relating to radioactive waste disposal: the Low-Level Radiation Waste Policy Act of 1980 and the Nuclear Waste Policy Act of 1982. At this time there are three disposal sites for commercially generated low-level waste in the United States: one located in Washington, one located in Nevada, and one in South Carolina (three other waste sites have been closed). Across the nation, *high-level* waste from nuclear power plants is maintained at plant sites in pools of water. Besides these acknowledged storage sites, there are at least thirty-five *unmarked* nuclear waste sites.

Public Law 97-425, the Nuclear Waste Policy Act, provides for the selection of two long-term repositories for high-level radioactive wastes. Attention is currently focusing on burying wastes deep underground. The first site, to be *ready* by 1998, will be located in Texas, Washington, or Nevada; the second will be *chosen* by 1998 from among twelve possibilities: three in Minnesota, two in Maine, two in North Carolina, two in Virginia, one in New Hampshire, one in Georgia and one in Wisconsin. The sum of $3 billion has been allotted for testing and drilling. A wise decision would be based primarily on geological considerations.

While there is an urgent need for long-term disposal, and even though specific proposals for underground storage are now being drawn up, the issue remains controversial. Additional proposals include burying wastes under the seabed or sending them into orbit around the earth or the moon. Despite all claims to the contrary, no one knows how to contain radioactive wastes for at least 10,000 years and keep them from contaminating the environment. This is the reason that most radioactive waste is still held in temporary facilities or dumped into landfills. Handling of waste has not been well controlled and examples of haphazard management abound. Leaks have been documented at almost all waste site locations.

The first leak of waste was discovered in 1956 at the government's 570-square-mile Hanford facility in Washington. It is estimated that about 450,000 gallons of high-level waste have leaked out of containers there since then. The Savannah River Plant, a military facility occupying 300 square miles near Aiken, South Carolina, has a terrible record for monitoring radioactive waste. A five-year study released by the Environmental Policy Group in July 1986 said that the degree of contamination at the plant is so severe that it is "a national sacrifice area." In an ABC-TV special in June 1985, Bill Lawless, a former senior engineer at Savannah River, recounted:

The liquid low-level wastes have for the past 32 years or so been dumped into seepage basins, which are just huge holes in the ground, and allowed to seep through the bottom of the basins into the groundwater system.

According to the Environmental Policy Group, these dumping procedures have already caught up with us: their report warns that the Savannah River waste threatens the Tuscaloosa aquifer, which supplies water for drinking and crop irrigation in South Carolina, Georgia, Florida and Alabama. In addition, the authors cite hazards posed by the presence of 51 tanks, containing a total of 27 million gallons of radioactive waste, at the plant. These tanks require continuous cooling to prevent them from boiling over. In the event of an earthquake or explosion, huge amounts of radiation would be released (see "Report Assails Safety of Nuclear Waste Storage at Carolina Plant," *The New York Times,* July 24, 1986).

Careless handling of the waste disposal problem has prevailed because no standards were set up by the Government until 1975. At that time, strict surveillance was imposed on commercial reactor waste. Nevertheless, the 1975 measures did not provide for monitoring of the wastes from weapons production. (The group that investigated conditions at Savannah River is a private, nonprofit organization.)

The Summer 1985 issue of *The Waste Paper,* a Sierra Club publication, revealed carelessness at the Department of Energy nuclear facility at Fernald, which is located in northwest Ohio, eighteen miles from Cincinnati. The plant refines uranium for military uses. The waste dump at Fernald contains about 500,000 tons of poison, some dating from the Manhattan Project (the development of the first atomic bomb). Plutonium, strontium and radium have leaked into local wells and drinking water—including the water of Fernald's Crosby Elementary School.

Similar extensive releases causing radiation exposures to nearby communities have been found at nuclear waste sites at West Valley, New York and Hanford, Washington. At West Valley (which has since been closed), the U.S. Geological Survey detected plutonium that had migrated sixty feet from the burial site. At virtually all storage areas for radioactive wastes, time has shown that what begins as an on-site problem eventually becomes an off-site problem as well.

The problems with nuclear plants don't stop once the plants cease operations. Reactors have a useful life of thirty to forty years. Older plants are already becoming lethal radiating structures. Even after the fuel rods are taken out, reactors remain intensely radioactive. Plants can either be dismantled and stored as waste or left to "cool down" for thirty to one hundred years and then dismantled. It is conceivable that the area for many miles around these "decommissioned" plants could become uninhabitable, prompting mass evacuations and relocations.

Dismantling plants will produce both low-level and high-level radioactive waste that needs a safe storage site for centuries. The remains of a single large reactor would fill more than a thousand trucks and need transportation to distant storage sites. Such transportation, of course, would add the risk of contamination being spread anywhere along the route.

Several experimental reactors and one small commercial reactor in Minnesota have been dismantled. The plant at Shipping port, Pennsylvania, which operated from 1957 to 1982, is in the process of being decommissioned. The Department of Energy has allotted $98 million for dismantling the reactor and shipping parts for storage at Hanford, Washington.

Nuclear reactors are licensed for forty years, although they usually fail by the time they are thirty. Eighty-one operating plants will close over the next few decades. By 2010, sixty-seven of those currently operating will reach the end of their licenses, and the rest by 2024. Unfortunately, however, thirty additional plants are scheduled to open soon.

As I stated at the beginning of this chapter—we are all very sick of hearing about pollution, but the fact remains that we are getting sick from pollution. In the next chapter, we will take a look at the ways in which low-level radiation damages our health.

2

How Radiation Damages Us

The mechanisms by which radiation harms our health are well worth understanding. Whether low-level radiation hits us from the outside air or is ingested in water or food, its method of harming us is the same: it disrupts at the level of the atom and molecule, causing imbalances that in turn can have vast ramifications.

The biological effects of radiation are determined by the interaction of energy and matter. One biological effect of radiation is *carcinogenesis,* which is the development of cancer. Another is the formation *oifree radicals.* Like radioactivity itself, these volatile and energetic pieces of matter can do a lot of damage to living cells and tissues. In this chapter, we will discuss the ways in which radiation damages health and take a look at the progress of research relating to radiation and health.

Radiation And The Blueprints For Our Cells

Radiation can have a big effect on our health. In order to understand how it causes such a big effect, it is important to think small. Living cells contain tremendous numbers of atoms and molecules, and the human body is made up of over ten trillion cells. When radiation strikes, damage to any *one* of the atoms in any *one* of the cells can have far-reaching consequences.

Our cells continually carry on metabolic processes such as taking in food, manufacturing proteins, getting rid of waste, and reproducing themselves. Every cell carries DNA, which is the blueprint for reproduction. When ionizing radiation hits a cell it can have several possible effects: it can kill it right away; it can harm the blueprint mechanism so that future cells are imperfect; it can damage the cell a certain amount so that it is repairable; or, it can pass by without causing injury.

A change in the genetic blueprint of a cell is known as a *mutation*. A mutation in a body cell can have many effects, including the development of cancer. An alteration in the blueprint of a sperm or egg cell can be passed on to the offspring. If the child is healthy enough to survive to adulthood, he or she can, in turn, pass the altered cell directions on to his or her children. Thus, the injury can persist through generations before manifesting itself. Geneticists point out this inherent hazard of radioactivity, emphasizing the possibilities of increased physical and mental deformities and higher incidence of disease as time goes on.

Free Radicals Impair Immunity

Free radicals can do a lot of damage. What are they, where do they come from, and what do they do? And—in practical terms—what can we do about them? If left unchecked, free radicals impair our immunity.

What are they? Stated very simply, *free radicals* are certain atoms or groups of atoms that are highly reactive because of the way their electrons behave. In general, there is a tendency for electrons to exist in pairs. However, odd or unpaired electrons are characteristic of free radicals. These electrons are easily dislodged from their orbits and can readily take part in chemical reactions. When electrons are knocked out of their orbits, they are referred to as *free electrons*. The important point about free radical reactions is that a seemingly small alteration can account for dramatic changes in biochemical properties. For example, if an ordinary molecule of oxygen picks up a free electron, it becomes the highly reactive free radical known as *superoxide*. Along with superoxide, other common substances involved in free radical reactions include hydrogen peroxide, hydroxyl and singlet oxygen. Free radicals can have either a positive, a negative or a neutral electrical charge. The vast majority of them are neutral (i.e., they have no charge).

Where do they come from? In small numbers, free radicals are a natural and normal part of existence. They are byproducts of various biochemical processes and the body normally keeps them in check. Exposure to ionizing radiation, however, greatly accelerates the formation of free radicals. (As mentioned in Chapter 1, ionizing radiation causes electrons to become dislodged from their orbits.) Chemical pollution exposure can also accelerate the formation of free radicals.

What do they do? Free radicals and the substances created as a result of free radical reactions can have a variety of devastating effects on our cells. If the number of free radicals is moderate, then the body can keep them in check. If there are too many, then there is overload and breakdown of the body's defense systems.

The key thing to be aware of is that the damage caused by free radicals multiplies rapidly. This is because they trigger a chain reaction: the formation of free radicals leads to the formation of *more* free radicals, each with the potential to disrupt our cells.

Free radicals can cause accumulations in the fluid between our cells (a factor that several authorities believe relates directly to aging). They can destroy the layer of fat in the cell membrane, which protects our cells. If they get turned loose inside the cell, free radicals can cause calcium imbalances and other problems. And finally, they can interfere with the cell's coding of genetic information.

Such interference can result in mistakes in the synthesis of proteins. This has serious consequences because we cannot live without the multitude of special proteins manufactured in the body. The synthesis of proteins is a vital, round-the-clock job. A protein that contains a mistake confuses the immune system. Under normal circumstances the immune system is able to distinguish "self" from "non-self." But it sees an abnormal protein as a foreign substance and tries to eliminate it. Eventually, waste material piles up faster than it can be handled by the lymph glands, which are the filters for the immune system. When the immune defense cannot keep up with the demands made upon it, viruses, microbes and cancer cells, which are ever-present—in fact, any kind of infection—can then take hold.

Counteracting Free Radicals

The way in which free radicals are normally kept in check is by the action of certain *free radical scavengers* that occur naturally in the body. These scavengers can donate an electron without becoming free radicals themselves. Certain enzymes serve this vital function. The body makes these as a matter of course. In addition, the work of these scavenger enzymes can be supplemented by a diet rich in *antioxidants* such as vitamins A, E and C and the mineral selenium. These nutrients are also scavengers, gobbling up free radical marauders—much like "Pac-Man" cleaning up the screen in the popular video game. Free radical scavengers are important components of the Diet for the Atomic Age.

The Progress Of Research Relating To Radiation And Health

Radiation has a life-shortening effect. It can cause cancer, leukemia, genetic damage, premature aging, abnormal blood clotting and a variety of other health problems. Radiation exposure has been linked to the development of arthritis, heart disease and diabetes, and can impair intellectual ability. Most insidious of all its effects, however, is the general lowering of immunity. Radiation depresses the body's ability to resist disease, and all sorts of nonspecific ailments (such as attacks of the "flu") result. The biological effects of radiation are so far-reaching that it hardly seems possible that they could be overlooked. Yet the reality of the destructive health effects of radiation emerged slowly.

In the 1950s, with enthusiasm for the "Atoms for Peace" program, facts were overlooked or swept under the rug. As mentioned in the previous chapter, until very recently, the United States Government repeatedly and systematically denied any connection between radiation exposure and long-term health problems among former servicemen, their families, and people living downwind of nuclear test sites.

In the 1950s and 1960s, people built fallout shelters not because they feared for their health but to shield themselves from the immediate effects of a possible nuclear weapons exchange. It was widely believed that a few days or weeks after a nuclear strike, families would be able to emerge from the shelters and resume their normal lives. But even thirty years ago, conscientious scientists and researchers around the world recognized the health hazards of the atomic age.

Looking back today, it is clear that these individuals set the stage for our growing awareness that all levels of radiation have long-term consequences in terms of human health. Although the antiwar and environmental movements of the 1960s and 1970s heightened public awareness of these problems, their impact is still not recognized by everyone. In the pages that follow, we will take a brief look at the progress of research relating to radiation and health, and at how the sobering facts first began to filter through to the general public.

Linus Pauling (who has won Nobel prizes for chemistry and for peace) was one of the few who spoke out throughout the fifties to warn the public about the genetic hazards of nuclear fallout. Saying that there was no safe threshold for radiation exposure, Pauling forecast that "bomb tests will ultimately produce about one million seriously defective children and about two million embryonic and neonatal defects." He noted: "Each nuclear test spreads the added burden of radioactive elements over every part of the world. Each added amount of radiation causes damage to the health of human beings all over the world and causes damage to the pool of human germ plasm such as to lead to an increase in the number of seriously defective children that will be born in future generations."

Dr. Alice Stewart, head of the Department of Preventive Medicine at Oxford University, England, and a respected epidemiologist (epidemiology is the study of public health), was one of the first to demonstrate the connection between low-level radiation and cancer. When she saw a large increase in the rate of leukemia (which is a type of cancer affecting white blood cells) among young children in England, she set up a study to find the cause. After researching the details relating to the death of 1300 children in the early 1950s, she found that babies born to mothers who had pelvic x-rays during pregnancy were twice as likely to develop cancer. She concluded that pelvic x-rays given to the mother in the first three months of pregnancy increased the risk of leukemia in the child. X-rays during the first three months were about sixteen times more carcinogenic than x-rays during the second or third trimester.

Although Dr. Stewart's warnings about the effects of radiation are now beginning to be recognized, with very few exceptions, thirty years ago her findings fell on deaf ears. In his book *Secret Fallout,* Dr. Ernest Sternglass, a nuclear physicist dedicated to alerting people to low-level radiation's effects, describes why information linking radiation exposure and health damage was ignored:

> If it were shown that large numbers of children had already died from the effects of fallout, then tremendous public revulsion would probably be generated against *all* activities that released more radioactivity into the environment. These would include not just the testing of nuclear weapons, but also the monumental program planned by many governments and industries throughout the world for the peacetime uses of atomic energy. For nuclear power reactors, atomic gas mining explosions, and other forms of nuclear engineering all normally release low levels of radioactivity.

In 1958, there were 64 aboveground tests in the United States and high amounts of strontium-90 were found in the milk of cows grazing downwind of the testing site. The first long-term study was published nearly thirty years later by Dr. Carl J. Johnson. It appeared in the conservative and widely respected *Journal of the American Medical Association* of January 13, 1984 with the title "Cancer Incidence in an Area of Radioactive Fallout Downwind From the Nevada Test Site." The study focused on Mormon residents of southwestern Utah, because they were noted for their healthful lifestyle and thus less prone to cancer. The researchers found higher than normal rates of leukemia along with high rates of other cancers, including cancers of the thyroid, gastrointestinal tract and breast.

Atmospheric nuclear weapons testing went on from 1951 through 1958 in the Western United States, mostly in Nevada. U.S. atmospheric testing of nuclear weapons also took place on islands in the central Pacific from 1946 to 1963. British nuclear tests were conducted in this area as well. The Nuclear Test Ban Treaty, which prohibited atmospheric bomb testing, was a 1963 agreement between the United States, the United Kingdom and the U.S.S.R. Nevertheless, other nations, such as France and China, were not bound by this treaty, and they continued aboveground testing until more recently. Underground testing in Nevada, the Pacific and other places around the world continues to the present day. This releases significant amounts of radioactivity into the environment—not only during the occasional mishap, but also under normal test conditions.

The controversy over radiation effects went on through the 1950s and 60s largely out of public view. The wisdom of the Test Ban Treaty was debated in 1963, but not many facts relating to health reached the public consciousness. In 1969, *Esquire* magazine published "The Death of All Children," one of the first public acknowledgements that fallout from bomb testing was much more serious than had been expected. In this article, Sternglass noted that recent studies comprised "the first documented, long-range analysis showing direct quantitative correlations between strontium-90 and infant mortality." In addition, he stated:

> The evidence available so far therefore suggests that radioactive strontium appears to be a far more serious hazard to man through its long-lasting action on the genetic material of the mammalian cell than had been expected on the basis of its well-known tendency to be incorporated into bone.

Sternglass has also investigated a correlation between prenatal radiation exposure and scholastic aptitude. He found that individuals born during the years in which large-scale nuclear testing prevailed had comparatively low test scores seventeen and eighteen years later, when they applied to college. As long as unborn children are exposed to radiation, there is the possibility of widespread intellectual decline. Sternglass poses this scenario:

> It would be the steady decline in the ability to read and reason and not so much the cancer in old age that would be the real seed for the self-destruction of a modern technological society.

New Breakthrough: Small Doses Have Cumulative Effects

Studies by Johnson, Gofman and Tamplin, Sternglass and others confirmed what was becoming accepted among all unbiased researchers: that radiation contaminates the environment and poses a long-term threat to human health. But in 1972 some *new* information came out. Scientist Dr. Abram Petkau, doing some research for the Canadian Atomic Energy Laboratories on how chemicals cross the cell membrane to enter the living cell, found that under some conditions the membranes broke much more readily than had previously been observed. Petkau found that it took a very large dose—3,500 rads (approximately equal to 35,000 years of background radiation)—to break the membranes when the dose was given by an x-ray machine in the course of a few minutes. One might assume that if the radiation was spread out over a period of hours or days, it would take the same total dose to break the cell membrane. But when Petkau irradiated the cells slowly (by adding a small amount of radioactive sodium salt to the water surrounding the cells) the cell membranes broke at a total dose of *less than one rad.* He varied the lengths of time that the cells were exposed to the sodium salts, and came to the conclusion that the longer the time of radiation exposure, the smaller the dose needed to do damage. With this observation, all assumptions as to the damage of small amounts of radiation were tossed up in the air.

Up to this time most opinions about low levels of radiation had been extrapolated from studies based on a higher dose for a shorter time. The two "data banks" that scientists depended upon were the health records of survivors of the Hiroshima blast and follow-ups on patients who had had radiation therapy. Sternglass comments on Petkau's research and its implications in *Secret Fallout:*

> It turned out that instead, a highly toxic, unstable form of ordinary oxygen normally found in cell fluids was created by the irradiation process, and that this so-called "free radical" was attracted to the cell membrane, where it initiated a chain reaction that gradually oxidized and thus weakened the molecules composing the membrane . . . .

Thus, almost overnight, the entire foundation of all existing assumptions as to the likely action of very low, protracted exposures as compared to short exposures at Hiroshima or even from brief, low-level medical X-rays had been shaken. Instead of a protracted or more gentle exposure being less harmful than a short flash, it turned out there were some conditions under which it could be the other way around: The low-level, low-rate exposure was more harmful to biological cells containing oxygen than the exposure given at a high rate or in a very brief moment. Petkau's finding that low-level exposure over a period of time is more harmful to body cells than the same exposure given at a high rate for a brief time was published in 1972 under the title "Effect of $22Na^+$ on a Phospholipid Membrane" in the journal *Health Physics.*

New Breakthrough: Some People Are More Susceptible

The breakthrough about small doses was followed and compounded by another new realization. Alice Stewart's study correlating prenatal x-ray exposure and childhood leukemia raised the possibility that some people might be more sensitive to radiation. While studying childhood leukemia and its relation to low-level radiation, Irwin Bross found that there is a wide range of sensitivity to radiation—age is a factor and previous medical history is a factor.

In the paper "Leukemia from Low Level Radiation: Identification of Susceptible Children," published in the prestigious *New England Journal of Medicine* in 1972, Bross proposed that receptivity to radiation "involved both the external hazard and the internal defenses of the subject exposed to the hazard."

We now know that infants and young children are especially susceptible to radiation damages, as are those with a lower level of health or a history of ailments. Interviewed in the book *Nuclear Witnesses,* Bertell stated: "Young adults with asthmas, severe allergies, heart disease, diabetes, arthritis and so on were about twelve times as susceptible to radiation-related leukemia as were healthy adults." In addition, she states that "Radiation effects are more pronounced when people who show signs of premature aging—like those with heart disease, diabetes, or signs of inability to cope with the environment, such as asthmas or allergies—are exposed."

Sternglass has explained how this new understanding that the healthy person is less susceptible to radiation effects destroys the value of earlier statistical studies:

> This factor alone would invalidate any estimate of the likely effect of small radiation exposures to a large human population, since these had been based on the average adult, obtained at high doses, and on the assumption of a linear relationship between dose and effect. For a non-homogeneous group, the more resistant individuals such as healthy young adults would not show any significant effects, while either the very young or the very old and those with immune deficiencies, allergies, and other special conditions might show an unexpectedly large effect.

Sternglass reviewed the evidence regarding variations in radiation susceptibility in testimony at Federal Court in Washington, DC in 1978. The book *Shutdown: Nuclear Power On Trial* transcribes the proceedings of that court. To Dr. Sternglass, all this evidence points "toward a single tragic conclusion"; that during the stage of early embryonic life, man is "hundreds of thousands of times more sensitive to radiation than anyone had ever suspected." Unfortunately, the radiation dose standards still in use today, which were calculated with Hiroshima survivors as a reference point, do not take differences in susceptibility into account.

Nuclear Workers Are in Danger

Another landmark study was inadvertently commissioned by the Atomic Energy Commission, in the late sixties, in an attempt to confirm the validity of the permissible or "safe" amounts allowed for nuclear workers. They asked Dr. Thomas Mancuso, a specialist in the field of occupational health, to study workers at four nuclear plants in the state of Washington.

Conducting the largest epidemiological study of radiation effects ever done, Mancuso found that workers had a 6 percent higher incidence of cancer and leukemia than the general population. But even more significantly, instead of confirming the accepted safety levels, Mancuso discovered that health effects occurred at an exposure level *as much as thirty times lower* than what was thought to be safe!

Clearly, there are serious implications not only for—nuclear workers, but for the general public as well. Over 90 percent of the U.S. population lives within 100 miles of a nuclear reactor. About one and a half million people in this country currently are exposed to low-level ionizing radiation in the workplace; over 30 million live within thirty miles of a nuclear power plant.

And the "safe" emissions from the nuclear plants in operation across America must be considered cumulatively, along with the effect of the total global fallout from nuclear weapons tests in the 1960s, 1970s and 1980s.

The results of Mancuso's long-range study were not appreciated by the Atomic Energy Commission and its successor, the Department of Energy. Mancuso's funding was cut off. An April 1979 article in the journal *Science* stated:

> DOE [the Department of Energy] withdrew its financial support in 1977 on the grounds that Mancuso's execution of the study was defective. Many people, including Mancuso, viewed the DOE action as an attempt to suppress findings unfavorable to the department's policies.

Mancuso worked with Alice Stewart, the epidemiologist who studied leukemia rates in England, and George Kneale, an expert statistician. After determining that the study was in fact valid, in 1977 they published the results in *Health Physics,* under the title "Radiation Exposures of Workers Dying from Cancer and Other Causes." As the authors of *Killing Our Own* observe:

> The impact of the new findings was hard to overstate. "The implications are far-reaching for health regulation and nuclear power in this country in general," said David

Auton, a physicist with the Defense Nuclear Agency. Standards for neutron radiation [one type of emission from nuclear plants] in particular might have to be tightened by a factor of ten and, on crucial jobs, the nuclear industry might have to hire ten times as many people [to reduce each individual's radiation exposure]. Exposure levels for people living near nuclear power plants would have to be reevaluated . . . . The new data, said Dr. Arthur Upton, former director of the National Cancer Institute, greatly strengthened the argument there is no "safe" level of exposure to radiation.

Another study of nuclear workers that confirmed higher cancer rates at lower than previously anticipated levels of exposure was done by a Boston blood specialist, Dr. Thomas Najarian. He was inspired by a leukemia patient who was a worker at the Portsmouth Naval Shipyard where atomic submarines are built and repaired. Najarian funded his own research and published his observations in the medical journal *Lancet* in 1978. He found that the shipyard workers had twice the cancer deaths and five times the leukemia deaths as the normal population. He observed that the workers probably received on the average about 200 millirems of radiation each year. This amount is about twice the amount of average background radiation that we receive, but is under the 500 millirems allowed annually for the general public. Nuclear workers are allowed up to 5,000 millirems a year.

It is possible that nuclear workers may exhibit premature aging due to radiation. In *Nuclear Witnesses,* Bertell recounts that in a 1979 visit to a plant that manufactured nuclear fuel rods, she spoke to a number of workers who showed signs of premature aging and had related health problems.

Everyday Dangers

It is the two recent discoveries—that those in a lower state of health are more sensitive to radiation, and that continuing low levels of radiation can do more damage than a single higher dose—which make it vital for us to recognize the everyday danger that has been heretofore ignored and misunderstood. Some of the everyday sources of radiation were examined in Chapter 1. In the pages that follow, we will take another look at the day-to-day impact of radiation on our external—and internal—environment.

Researchers from Heidelberg University conducted lengthy studies for the planned Wyhl nuclear reactor in Germany. In addition to confirming that some members of the population are more susceptible to the effects of radiation than others, they found that the amount of radioactive substances that actually go from the power plant to the human had been underestimated by a factor of 100 to 1000. This miscalculation resulted from a failure to consider the nature of the *food chain.*

What Is The Food Chain?

The food chain starts with green plants, which absorb energy from the sun and store it. Animals, which cannot obtain energy directly from the sun, need to eat large numbers of plants (or to eat other animals that have eaten plants) in order to meet their energy requirements. Unfortunately, not only usable energy is passed along the food chain—contaminants are also passed along, from plants to animals to man.

The authors of the 1979 Heidelberg Report found that this concentration had not been correctly taken into account. Nor had the fact that a radioactive substance, when complexed (chemically combined) with another substance, can have an amplified effect. For example, plutonium that gets into our drinking water has one value when measured alone, but when it is complexed with chlorine, its potency can be increased by up to one thousand times.

The Heidelberg Report contended that estimates of the amounts of plutonium, cesium and strontium picked up by vegetation may be up to one thousand times too low. The fact is that even if the concentrations of radiation in releases from power plants are within the acceptable levels, the materials can become concentrated in the food chain.

It has been found that fish, for instance, that feed on plankton and algae and ocean bottom sediments concentrate radionuclides to levels vastly exceeding the amounts in the water they live in. The term *transfer factor* (or *concentration factor)* is used to measure this increase. Various studies have found that the transfer factor for plutonium in an algae called *Sargassum* is 20,000; the transfer factor for strontium-90 in clam shells can range up to 65,000. A study on ducks around the Hanford, Washington plants found the cesium-137 transfer factor to be from 2,000 to 2,500. Cesium-137, which is emitted daily from nuclear plants, tends to concentrate in muscle tissue.

The transfer factor reflects the stages the radioactive substance goes through after being released from the plant; from air, to soil, to plant, to meat, to intestinal tract absorption by the human. The links in the food chain are far from simple—a contaminant such as cesium-137 can take any of several alternate pathways to the human body (see the illustration on page 52). Cesium-137 may go out with liquid waste into a nearby river; from there it can go into the underground water table or be used to irrigate crops. The water table can contaminate drinking water; crops, of course, will eventually be harvested and

Routes of Contamination from Nuclear Plant Emissions to Us

used for food. Crops may either be consumed directly or used to feed livestock—which will then yield contaminated milk and meat.

The following chart lists the dose of radiation that various foods supply according to the pathway of contamination, whether by air or by water. The closes are based on the indicated estimated average annual consumption (by adults) for those foods. It indicates that, in general, foods that are lower on the food chain have less radiation contamination than those higher on the food chain (a few exceptions to this rule will be discussed.

Whole Body Doses from Various Foods

Radiation Exposure by Exhaust Air Pathway

Foodstuff	Millirems per year	Estimated Annual Consumption
Leafy vegetables	10.7	110 pounds
Grains	13.1	198 pounds
Root vegetables	39.8	110 pounds
Potatoes	128.7	198 pounds
Milk	162.1	382 quarts
beef	348.9	220 pounds

Radiation Exposure by Waste Water Pathway

Foodstuff	Millirems per year	Estimated Annual Consumption
Leafy vegetables	4.9	110 pounds
Milk	8.9	382 quarts
beef	14.6	220 pounds
Root vegetables	24.8	110 pounds
fish	71.6*	110 pounds

*Fish from the cooling water outlet had a dose value of 117.3 mrem/yr.

The food chain is affected by nuclear waste products at all levels, and concentrations are higher depending on how many stages it has gone through. When we say a food is "high on the food chain," we mean that it contains substances that have gone through several stages and have become more concentrated at each stage. Cesium, which has a half-life of thirty years, is active for six hundred years. Today's daily nuclear plant emissions accumulate in the environment on top of yesterday's plant emissions, and tomorrow's will accumulate on top of today's.

As the scientists with the Department of Environmental Protection at Heidelberg University point out in the abstract to their report:

> The often stated value of 1 mrem/year [1 millirem per year] radiation dose to the public from radioactive emissions from nuclear power stations is the result of calculations and not of measurements. Even in routine releases nuclear power reactors emit hundreds of radionuclides . . . . Radionuclides which are discharged into the environment, undergo a great number of transport processes, where they are more or less diluted or enriched and can lead by many different ways to radiation exposure of the individual.

They also state: "As may be seen, the recommended values may in special cases underestimate the transfer of radionuclides 10-fold, 100-fold, or even 1000-fold."

Since 1956, the Sellafield (previously called Windscale) reprocessing plant in England has released—on a daily basis—about 1.2 million gallons of liquid waste including cesium, strontium and plutonium into the Irish Sea. The plutonium becomes concentrated in the sea water. Droplets of water at the shoreline were found to be 800 times more radioactive than the rest of the seawater.

In 1983, the plant's owners were convicted of criminal negligence and fined $15,000. The effects of their negligence, however, are still being felt by the people who live nearby. One farmer whose land adjoins the plant has lost forty cows. Fish from the Irish Sea and the black-headed gulls that feed on them have also suffered. The gulls, valued for their eggs, have begun to die off. In a recent *Newsday* article ("Nuclear Time Bomb," May 20, 1986), reporter Patrick J. Sloyan described how these changes have affected the local population:

> Gone . . . are the lines of housewives who rushed to the shoreline each day to buy the cod caught by Paul Peterson and Andrew Graham in the deep holes of the Irish Sea. But the cod today is full of radioactive cesium from the Sellafield pipeline. So are the winkles, the clams and other seafood.
>
> Some of the fish eaters in the area already had more than half the cesium level considered safe to absorb in a lifetime.

Chernobyl and the Food Chain

The nature of the dangers posed by everyday nuclear power plant emissions have been highlighted by the explosion and fire at the Chernobyl complex in the U.S.S.R. on April 26, 1986. Radioactive fallout, in the form of particles and gases, was carried across international borders by the wind. In addition to the vast areas in the Soviet Union that were contaminated, parts of Rumania, Poland, East and West Germany, Austria, Hungary, Yugoslavia, Czechoslovakia, Greece, Turkey and Switzerland were affected. Radioactive plumes also extended north to Sweden, Finland and Norway. Contamination even stretched into portions of England, France and Italy.

Upper air currents carried the radioactive cloud to Korea, Japan, the Pacific islands and across the United States as well. Some weeks after the Chernobyl accident, fallout was detected in air, milk and water samples in this country. It is interesting to view the accident with reference to the food chain.

The Lawrence Livermore National Laboratory has estimated that about 40 percent of the Chernobyl reactor's radioactive fission products were ejected in the initial explosion and fire, and another 10 percent were released over the following few days. The release of radioactive cesium amounted to about 40 million curies. Other authorities estimate that the reactor contained about 1,000 pounds of plutonium. Although the Soviets have claimed that only 3 to 5 percent of the core material was lost, in their report to the International Atomic Energy Association they acknowledged a release amounting to 50 million curies of gaseous material and 50 million curies in the form of particles.

Because the initial explosion was so energetic and the graphite fire created a chimney-like updraft, it is believed that a large part of the radioactive debris was carried very high into the atmosphere. If this was so, then the levels of fallout close to the plant soon after the tragic accident might have been low—and misleadingly so. Later repercussions may very well include the appearance of long-term products in root vegetables not only in the fall crop but in all types of crops for years to come.

One interesting point regarding Chernobyl and the food chain is the misconception that most media accounts of this accident perpetrated. Most stated that the radioactive material released from Chernobyl dissipated as it traveled away from the source—portraying the problem as one mainly of geography. Such discussions left out one other relevant point—the fact that the concentration of radioactive material increases as it goes up the food chain.

For instance, many newspapers and magazines printed in May 1986 noted that people living in an area up to 60 miles away from the accident site could be expected to have a significant increase in cancer over the next decades, but that for those beyond a 200-mile radius the risks were quite small. In contrast, according to Dr. Ernest Sternglass, even 500 picocuries of iodine—an amount found in rainwater in the United States following the accident—could cause serious damage to a developing fetus or infant.

Even though several nations issued warnings to guard against the consumption of milk, meat, fresh vegetables and fruit, the popular press described the health hazards in a way that ignored the valid and vital food chain concept.

By and large, the long-term potential for concentration as radionuclides circulate through the atmosphere, rainfall, soil and crops is not given its public due. This is a grievous error if we are to protect ourselves from the negative health effects of radiation. Patterns of food imports and exports, along with wildfowl migration, further complicate the picture. In the aftermath of Chernobyl, many individuals may have unwittingly consumed food which was grown within the 200-mile "danger" limit or was contaminated despite being located at a seemingly great distance from the source.

Despite the mass media's downplaying of the food problems arising from the Chernobyl catastrophe, the Soviet Union took a number of steps to protect its citizens. Livestock within at least 20 miles of the Chernobyl plant was slaughtered. Acres of grazing land were abandoned because they were too radioactive. The Soviet Union now plans to decontaminate 1,000 square miles of farmland surrounding Chernobyl. (Decontamination procedures are still somewhat experimental; their long-term effectiveness remains to be seen. The techniques used in the U.S.S.R. include entombing the damaged reactor in concrete; pulling up trees, shrubs and crops, and burying them like radioactive waste; sealing old wells and digging new deep wells; building dikes to prevent water contaminated by radiation from entering rivers; dumping mixed materials such as lead, limestone, sand and boron over affected areas; paving roads; and washing buildings.)

Other nations responded to the crisis as well. In May 1986, the European Economic Community issued temporary restrictions on imports of fruit, vegetables, milk and meat originating from within a 625-mile radius of the plant. The ban affected foodstuffs from the U.S.S.R., Hungary, Poland, Rumania, Czechoslovakia, Austria and Yugoslavia. (But since fallout normally occurs where there is rainfall and not within a set radius of a nuclear contamination site, this action may have been more political than rational. Radionuclides deposited in clouds can be carried great distances—more than 625 miles from their point of origin.)

In Italy there was a ban on the sale of leafy vegetables and the government advised against the consumption of milk for several weeks in May. In West Germany, milk and milk products were banned and the public was advised not to eat lettuce and spinach until mid-June. Various other European countries issued warnings against drinking rainwater or milk and eating fresh surface vegetables. Up to 1,000 miles away from the accident site, such precautions were vitally important not only to prevent immediate radiocontamination, but also to help prevent unsafe concentrations of radioactive elements from accumulating along the food chain.

In Sweden, however, condemned milk was spread over farmland. The hay, contaminated with cesium and ruthenium, was harvested for winter fodder! This is sure to have negative ecological impact.

The principle of selective uptake, which we will examine in the following chapter, suggests two promising methods of reducing the impact that long-lived radionuclides will have on the food chain. Calcium compounds applied to the soil may bind with strontium-90, preventing its uptake by growing plants. Similarly, the use of potassium-rich fertilizers can keep plants from absorbing harmful cesium.

THERE IS NO SAFE DOSE OF RADIATION

In the early days of radiation research the scientific authorities assumed that there was a dose of radiation which was "safe," and the false illusion of a safe amount of radiation pervaded those active in the promotion of nuclear power. The chart below summarizes the current radiation dose limits that have been recommended by the National Council on Radiation Protection (NCRP). Standards are set by the Environmental Protection Agency (EPA).

Radiation Dose Limits Recommended by the National Council on Radiation Protection

Group or Situation	Dose Limit
radiation workers	5 rems per year
general public	0.17 rem per year average
	.5 rem per year should not be exceeded

Exceptions are made in the event of an accidental radiation release; in these circumstances "acceptable" doses are much higher.

A 25-rem limit for radiation workers was in effect in the United States through the 1940s. The Atomic Energy Commission (AEC) reported in 1949:

> Many years ago medical scientists agreed that a normal human being could sustain a continuous day-in-and-day-out whole-body external exposure of one-tenth roentgen per day (25 rems a year) without any detectable effects.

One year later the dose was reduced to 15 rems a year and the commission published a report stating:

> Through long study of the effects of such exposures it has been determined that a dose of 0.3 roentgen per week (15 rems a year) may be delivered to the whole body for an indefinite period without hazard.

Six years later (in 1956), the government cut the allowable dose for radiation workers to 5 rems a year, where it remains today. The following chart shows the progressive reduction of allowable radiation exposure standards. The first recommendations were set in 1925 after it became apparent that unlimited exposure was unsafe. Standards for amounts referred to x-rays and were measured in roentgens (R). In 1949, the measure was changed to rems and was based on amounts of radiation from bomb development and nuclear plant emanations. Standards were reduced with more understanding of dose/effects but still allowing industry's risk/benefit.

Changes in Levels of Permissible Exposure to Radiation

For Radiation Workers

Year	Recommended Maximum
1925	52 R per year
1934	36 R per year
1950	15 rem per year
1957	5 rem per year

For the General Public

1952	1.5 rem per year
1958	0.5 rem per year (maximum dose for overall population)
1958	0.17 rem per year (average for individual)

R = roentgen rem = roentgen equivalent man 1R = .88 rem

Source: Karl J. Morgan, "Cancer and Low-Level Ionizing Radiation," *Bulletin of the Atomic Scientists,* September 1978.

By the mid-1960s, the United States Government had agreed that the previously assumed idea of a safe threshold was invalid and that it was impossible to set a safe level for individual exposure. The Government recognized (at least in theory) that any dose, no matter how small or of what duration, is capable of causing adverse health effects. For instance, a report issued in 1966 by the Atomic Energy Commission acknowledged radiation's potential, yet unknown, effects upon the human gene:

> No matter how small a dosage of radiation the gonads [reproductive organs] receive this will be reflected in a proportionately increased likelihood of mutated sex cells with effects that will show up in succeeding generations . . . . That is why there is no threshold in the genetic effect of radiation and why there is no safe amount of radiation insofar as genetic effects are concerned. However small the quantity of radiation absorbed mankind must be prepared to pay the price in corresponding increase in genetic load.

The 1966 AEC report "The Genetic Effects of Radiation" said:

> If the number of those affected is increased there would come a crucial point, or threshold, where the slack could no longer be taken up (by those not affected). The genetic load might increase to the point where the species *as a*

whole would degenerate and fade toward extinction—a
sort of 'racial radiation sickness' [italics added].

These acknowledgements aside, the AEC went ahead and set
a number amount on how many rems or rads were acceptable per
person per year, without stating that this amount causes some of the
population certain health damage. Gofman explains the government's
perception of the relationship between risk and benefit as follows:

> What the standard-recommending bodies hoped was that
> the benefits of the atomic technology might offset the
> cancer, leukemia and genetic hazards.

And he goes on to say what it is he thinks we are actually
doing:

> Licensing a nuclear power plant is, in my view, licensing
> random, premeditated murder . . . . They allow workers
> to get irradiated, and they have an allowable dose for the
> population. So, in essence, I can figure out from their
> allowable amounts how many they are willing to kill per
> year.

In 1958, the dose amount for the general public was reduced
from 1.5 rems per year, an amount set in 1952, to 0.17 rem per
year. This amount was arrived at because it is roughly equal to the
amount of radiation from natural and medical sources that a person
in the United States receives in a year. In effect, the 0.17 rem level
permits an individual to receive *double* the background amount of
radiation exposure. Doubling the naturally occurring amount of
radiation could double the present incidence of disease and increase
the amount of genetic damage. This could lead to a vastly increased
number of births of defective children.

Dr. Bertell agrees:

> I would say that in the present state of the art the entire
> nuclear industry constitutes one colossal experiment on
> humankind. It is unjustifiable either in terms of energy

needs or national defense. We are already overcommitted
to this life-threatening technology, the consequences of
which extend into the next hundred thousand years. It is
time to develop a whole new set of values and priorities,
time to develop a whole new approach to energy problems
and ways to ensure the peace and well-being of all life on
the planet.

When researching health effects of x-rays, Bertell was shocked
by the levels of radiation exposure that are accepted. She says: "I
had been measuring the health effects of one, two, three and four
and five chest x-rays. Then I found the Federal Government allows
the general public to receive up to five hundred millirems per year.
Moreover, I learned that nuclear workers are allowed to receive
up to five rems—which is the bone marrow equivalent of one
thousand x-rays per year. These are the federal regulation protection
standards!"

Helen Caldicott, a medical doctor who gave up her practice in
pediatrics to use all her time to warn the public about the dangers
of low-level radiation, does just that in her book *Nuclear Madness:*
"The truth is that we are courting catastrophe. The permissive
radiation policy supported by the American government in effect
turns us into guinea pigs in an experiment to determine how much
radioactive material can be released into the environment before
major epidemics of cancer, leukemia, and genetic abnormalities take
their toll."

In an article in *The Nation,* Harvey Wasserman and Norman
Solomon (the authors of the book *Killing Our Own)* put the problem
in perspective: "The evidence is overwhelming that thousands
of Americans exposed to what they were told were 'safe' doses
of low level radiation, are now suffering from cancer and other
radiation-related diseases. Their story, from Hiroshima to Three
Mile Island and beyond, may well be the biggest health scandal in
United States history."

Nevertheless, early in 1986, the Nuclear Regulatory Commission
was proposing a rule to *increase*—by up to ten times or more—the
current "acceptable" limits for radiation exposure to the public,
to workers, and to the environment. The NRC proposal, entitled

"Part II, Title 10, Parts 19 et al, Standards for Protection Against Radiation," was published in the *Code of Federal Regulations* in January 1986.

Its provisions include: increasing the permissible exposure of 65 percent of the most significant radionuclides (including strontium-90 and iodine-131) up to ten times above the present levels; and permitting the doubling of workers' radiation exposures without the consent of the workers. In addition, under the proposed rule, small amounts ("de minimus") of radioactive releases may be exempt from being monitored. A controversial way of recalculating radiation doses and their health effects, first proposed in 1977, underlies the NRC proposal. To many observers, it seems clear that these newly proposed "permissible" levels are actually designed to foster the expansion of the nuclear industry, rather than to protect human health.

On the international level, various recommendations have been made, but individual countries have the ultimate control over their own radiation dosage standards. Some are stricter than ours, others are less strict. All of the regulatory agencies, however, base exposure limits on calculated safety levels. As we have seen, there is no safety threshold—individuals vary in their susceptibility to radiation damage, and the dose levels that cause damage have been grossly underestimated.

There is no safe dose of radiation, but we can take steps to improve our overall health and reduce our susceptibility to radiation's potentially devastating effects.

3

Survival of the Fittest

The highest possible state of health is the best security for atomic age. I remember my grandmother insisting on the importance of a healthy diet, but I didn't pay too much attentior then. Now it is imperative. The Radiation Shield plan based on two principles: the principle of selective uptake and the principle of the healthy survivor. In this chapter, we will examine these ideas. In subsequent chapters, we will discuss their practical applications.

The Principle Of Selective Uptake

Radioactive elements often behave like similar non-radioactive elements. The principle of *selective uptake* is based on the demonstrable biological fact that when our cells are saturated with the nutrients they need, there is less chance for radioactive elements to move in. But if our diets are deficient in vital nutrients such as calcium and potassium for an extended period of time, our uptake of their radioactive counterparts is increased.

Similar Elements Have Similar Properties

Many elements are quite similar in structure to one another. Elements are placed into groups in the *periodic table* (see page 306) determined by the number of electrons orbiting in the outermost shell of their atomic structure. These outermost electrons (which are known as *valence* electrons) enter into chemical reactions. Elements within each grouping (or "period," or, one could say, "family") react chemically in similar ways.

In place of the specific element it actually needs, the human body may pick up a "similar" element in the same group. This is central to the concept of selective uptake. Below you will find a listing of the elements in groups corresponding to the periodic table. This shows which elements the body may use when the needed nutrients are not available.

The body's survival mechanism tends to grab the things we need for normal functioning and health. When an element is lacking, the body takes up a similar element that is available. For example, strontium is quite similar in structure to calcium (they are in the same group in the periodic table). Strontium-90, a radionuclide that is found in all nuclear fallout and power plant emissions, and calcium can enter into the same types of chemical reactions. Thus, the human body uses them in the same way, primarily to strengthen bones and teeth.

Fortunately, however, the body is selective. Normal (that is, non-radioactive) elements have priority. When non-radioactive elements are available, the body tends to not absorb the radioactive ones. The situation can be likened to that of a baseball team that has runners on base. The team is more likely to come out ahead when the bases are loaded—it's better to have runners on first, second and third than it is to leave any bases unoccupied. If all of the positions are already occupied by calcium then strontium will be prevented from going there. Deficiencies are what lead to the uptake of radioactive elements.

Groups of Similar Elements
Based on Number of Valence Electrons

1. Group IA	2. Group IIA	3. Group IIIA
hydrogen (H)	beryllium (Be)	boron (B)
lithium (Li)	magnesium (Mg)	aluminum (Al)
sodium (Na)	calcium (Ca)	gallium (Ga)
potassium (K)	strontium (Sr)	indium (In)
rubidium (Rb)	barium (Ba)	thallium (Tl)
cesium (Cs)	radium (Ra)	
francium (Fr)		

4. Group IVA
carbon (C)
silicon (Si)
germanium (Ge)
tin (Sn)
lead (Pb)

5. Group VA
nitrogen (N)
phosphorus (P)
arsenic (As)
antimony (Sb)
bismuth (Bi)

6. Group VIA
oxygen (O)
sulfur (S)
selenium (Se)
tellurium (Te)
polonium (Po)

7. Group VIIA
fluorine (F)
chlorine (Cl)
bromine (Br)
iodine (I)
astatine (At)

8. Group VHIA
iron (Fe)
ruthenium (Ru)
osmium (Os)

9. cobalt (Co)
rhodium (Rh)
iridium (Ir)
unnilennium (Une)

10. nickel (Ni)
palladium (Pd)
platinum (Pt)

11. Group IB
copper (Cu)
silver (Ag)
gold (Au)

12. Group IIB
zinc (Zn)
cadmium (Cd)
mercury (Hg)

13. Group IHB
scandium (Sc)
yttrium (Y)
lanthanum (La)
actinium (Ac)

14. Group IVB
titanium (Ti)
zirconium (Zr)
hafnium (Hf)
unnilquadium (Unq)

15. Group VB
vanadium (V)
niobium (Nb)
tantalum (Ta)
unnilpentium (Unp)

16. Group VIB
chromium (Cr)
molybdenum (Mo)
tungsten [wolfram] (W)
unnilhexium (Unh)

17. Group VIIB
manganese (Mn)
technetium (Tc)
rhenium (Re)
unnilseptium (Uns)

18. Group 0
helium (He)
neon (Ne)
argon (Ar)
krypton (Kr)
xenon (Xe)
radon (Rn)

19. Lanthanide series
lanthanum (La)
cerium (Ce)
praseodymium (Pr)
neodymium (Nd)
promethium (Pm)
samarium (Sm)
europium (Eu)
gadolinium (Gd)
terbium (Tb)
dysprosium (Dy)
holmium (Ho)
erbium (Er)
thulium (Tm)
ytterbium (Yb)
lutetium (Lu)

20. Actinide series
actinium (Ac)
thorium (Th)
protactinium (Pa)
uranium (U)
neptunium (Np)
plutonium (Pu)
americium (Am)
curium (Cm)
berkelium (Bk)
californium (Cf)
einsteinium (Es)
fermium (Fm)
mendelevium (Md)
nobelium (No)
lawrencium (Lr)

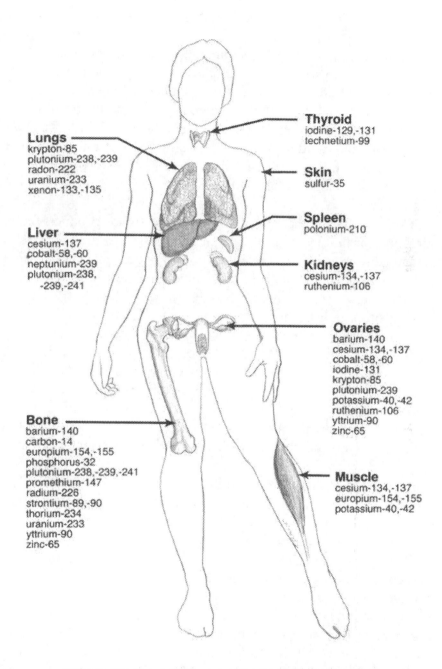

Thyroid
iodine-129,-131
technetium-99

Lungs
krypton-85
plutonium-238,-239
radon-222
uranium-233
xenon-133,-135

Skin
sulfur-35

Spleen
polonium-210

Liver
cesium-137
cobalt-58,-60
neptunium-239
plutonium-238,
   -239,-241

Kidneys
cesium-134,-137
ruthenium-106

Ovaries
barium-140
cesium-134,-137
cobalt-58,-60
iodine-131
krypton-85
plutonium-239
potassium-40,-42
ruthenium-106
yttrium-90
zinc-65

Bone
barium-140
carbon-14
europium-154,-155
phosphorus-32
plutonium-238,-239,-241
promethium-147
radium-226
strontium-89,-90
thorium-234
uranium-233
yttrium-90
zinc-65

Muscle
cesium-134,-137
europium-154,-155
potassium-40,-42

Where Radionuclides are Deposited in the Body

Selective Uptake and Specific Organs

When radioactive substances are taken up by the body, they tend to accumulate in various tissues and organs, as depicted on page 68. We have seen that if there is not enough calcium in the body, and calcium is not readily available, the body will pick up strontium-90 (or other radioactive elements that are similar to calcium—such as barium-140 and radium-226) and deposit it in the teeth and bones. Once deposited, strontium irradiates nearby cells. It can reduce normal function immediately, and lead to the development of cancer in the future.

Cesium-137, another one of the long-lived fission products found in all nuclear fallout and power plant emissions, is a radionuclide in the same group as potassium. Taken up by the body when potassium is lacking, cesium becomes concentrated in the muscles and reproductive organs, particularly the ovaries. (Radioactive forms of potassium, potassium-40 and potassium-42, also concentrate in these areas of the body.)

In April 1974, the journal *Health Physics* published an article that noted: "detailed information on cesium transport is urgently required in connection with incorporation and decorporation processes in a biosphere contaminated with radiocesium." Contamination of the biosphere is no small matter. The biosphere encompasses our entire environment on the earth's surface—the land, air and water that support all living things.

Iodine is a nutrient that is utilized by the thyroid gland in the formation of certain hormones. These iodine-containing hormones are important in regulating body processes. If the amount of iodine in the diet is inadequate, the body will take in radioactive iodine-131 to fill its need and saturate the thyroid. Radioactive iodine can be inhaled (in gaseous form) or ingested (in the form of iodine salts). In either case, once the iodine-131 migrates to the thyroid gland, it will irradiate nearby cells, and depending on the amount present, will impair thyroid functions and/or cause cancer later on. In *Nuclear Madness,* Dr. Helen Caldicott states that the time between the thyroid's uptake of iodine-131 and the development of cancer may be "twelve to fifty years." Other serious problems arise when iodine-131 accumulates in the thyroid of the developing fetus. This leads to a reduced rate of growth, low birth weight, and increased infant mortality.

A report published in August 1977 by the National Council on Radiation Protection and Measurements (NCRP Report No. 55, "Protection of the Thyroid Gland in the Event of Release of Radioiodine") suggested that potassium iodide be considered for "thyroid-blocking" in an emergency. The principle of selective uptake explains this proposal. The non-radioactive iodine (contained in the potassium iodide compound) would be taken up by the thyroid gland to such an extent that the thyroid would be saturated—unable to absorb any more iodine. Thus, the body's uptake of radioactive iodine would be blocked.

The 1977 NCRP report also proposed stockpiling iodine compounds for use in the event of a radiation emergency. Shortly after the nuclear accidents at Three Mile Island in 1979 and Chernobyl in 1986, efforts were made to distribute iodine preparations such as potassium iodide in areas where the population was exposed to radiation. This was the reason—to try to limit the absorption of iodine-131. Of course, this is only a partial solution to the health hazards created by nuclear fallout, because it only protects people from that one element.

Yet I have not heard of any large-scale plans to distribute blocking agents for other radioactive elements. If this principle is acknowledged to work for one set of radioactive and stable elements, I wonder why it has not been applied to other groups of elements.

Plutonium, which has a half-life of thousands of years, is a radioactive element, similar in structure to iron, that can be picked up in the human bloodstream and carried to iron storage cells in the liver and bone marrow, where it stays, as the other radionuclides do, to irradiate the surrounding cells.

In addition, according to Caldicott: "Plutonium's iron-like properties also permit the element to cross the highly selective placental barrier and reach the developing fetus, possibly causing . . . damage and subsequent gross deformities in the newborn infant."

Radioactive zinc-65 can be incorporated in the body instead of the stable forms of this mineral; it is deposited in the bones and reproductive organs. Cobalt-60 can be used instead of stable forms of cobalt by microorganisms manufacturing vitamin B_{12}. Vitamin B_{12}

is a complex compound that functions in the synthesis of red blood cells and DNA. In addition, it is important in maintaining the health of the nervous system. The body stores vitamin B_{12} for long periods of time. Adequate reserves of B_{12} can block the body's uptake of B_{12} containing radioactive cobalt. A listing of the stable elements that block the uptake of radioactive ones appears below.

Selective Uptake

Stable Element	Radioactive Element
Calcium	Strontium-90
Iodine	Iodine-131
Iron	Plutonium-238, -239
Potassium	Cesium-137
Sulfur	Sulfur-35
Vitamin B_{12}*	Cobalt-60
Zinc	Zinc-65

The stable elements listed in the left-hand column block the uptake of the radioactive elements listed in the right-hand column. (The stable isotopes of iodine, sulfur and zinc are iodine-127, sulfur-32 and zinc-64.)

* Vitamin B_{12} is a compound containing the element cobalt. (Stable cobalt is cobalt-59.)

How Essential Minerals Protect the Body

It seems clear that the body can reduce absorption of radioactive elements by building up reserves of the essential minerals in which it is low. Many investigators have studied the blocking effects of adequate iodine and calcium intake. In various ways, these essential minerals can reduce the body's utilization of radioactive substances.

Health researcher Dr. Richard Passwater notes that "Normal dietary iodine can protect against the absorption of the radioactive iodine. Inhaled radioactive gas or ingested radioactive iodine salts will be rapidly excreted if the thyroid is in no need of iodine."

In an article in a collection of papers dealing with chemical and radionuclide food contamination, published in 1972, M. E. Alpert et al. write:

There have been several studies on prophylactic [preventive] and therapeutic measures to reduce the dose subsequent to human intake of radioiodine. The effect of these measures is based on reducing the uptake to the thyroid and/or the effective half-life. . . . If a blocking dose, about 100 mg of stable iodine, is given within about 2 hours or less of initial iodine-131 intake, the percent uptake to the thyroid is reduced by about 90%.

But, wouldn't you say, given what we know, that it is a good idea to *always* have the thyroid's iodine needs met? In Japan, where the diet includes ocean fish and seaweeds with high amounts of natural iodine, people are in fact doing this—and Japan has the lowest infant mortality rate in the world. The Japanese diet is protective, as it leaves little room for the uptake of radioactive iodine.

A study published in the 1967 book *Strontium Metabolism,* edited by J. Lenihan, found that: "Addition of Ca [calcium] to the diet of experimental human subjects reduced the retention and accumulation of Sr90 [strontium-90] in bone."

In the papers on chemical radioactive food contamination mentioned above, M. E. Alpert et al. observe: "Newly formed bone has about the same strontium-calcium ratio as is in the blood circulating at the time of formation. There is some discrimination against strontium which may be influenced by preferential absorption of calcium through the gut [digestive tract]." The researchers found that if dietary calcium was available, the body was four times more likely to utilize it in the formation of bone than it was to use strontium.

As reported in a 1967 issue of the *International Journal of Applied Radiation and Isotopes,* doctors at the Veterans Administration Hospital in Hines, Illinois investigated the effects of calcium and magnesium on the body's intake of radioactive strontium. They found that when calcium and magnesium figured in the diet, a significant amount of radiostrontium was removed from the bloodstream and excreted.

I haven't found in the medical literature many experiments with humans on the specific blocking power of adequate mineral nutrition. Medical practice doesn't usually allow any toxic substance to be given to humans for research, so demonstrations of the radioprotective benefits of adequate mineral nutrition in humans are limited. However, we can conclude from research done with iodine-131 and strontium-90 that other radioactive elements such as cesium-137, zinc-65 and plutonium are blocked from being taken up by the body by their essential counterparts of potassium, iron and zinc.

The Healthy Survivor Theory

As mentioned in Chapter 2, some groups in any given population are more susceptible to radiation. These groups include: infants in utero, babies and young children, because their cells are dividing rapidly; older people, because they generally have weakened immune systems due to having already been exposed to a cumulative amount of background and man-made radiation; and everyone who is in a weaker state of health. The *healthy survivor theory,* which was formulated by the British epidemiologist Alice Stewart, is based on scientific observations that those who managed to survive the early infections and diseases that set in following radiation exposure were less likely to die of cancer later on. They may very well have been healthier to begin with. Put simply—it seems clear that people who enjoy optimal levels of health are the most resistant to radiation damage.

You may recall from our earlier discussion that the concept of the direct relation between radiation doses and health effects has been revised as radiation research has progressed. Once it was recognized that lower doses could do more damage, a major misconception was clarified—namely, that the biological system receiving the dose is the only factor determining the extent of the damage. But for our needs now, it is a crucial factor. We can influence the amount of radioactive elements that reach the cells of the immune system.

A Landmark Study

Dr. Alice Stewart's landmark study, "Delayed Effects of A-Bomb Radiation: A Review of Recent Mortality Rates and Risk Estimates for Five Year Survivors," was published in the *Journal of Epidemiology and Community Health* in June 1982. Stewart's study provides evidence that those who are the healthiest are most likely to survive radiation exposure.

Stewart observed the healthy survivor effect at Hiroshima—that is, the individuals who did not succumb immediately to radiation from bomb fallout were healthier than their neighbors who did die shortly after the atomic blast. She made this determination by comparing their health statistics in the years after the bombing to those of the Japanese population in general.

She studied the death rate of the survivors in the years since 1945, and for certain diseases that are not considered to be affected by radiation, she found the death rate was much lower than among the general population. She suggests that many studies that had been done on these survivors are misleading, and that all analysis of fallout effects on those near where the bomb fell should be adjusted for the fact of the healthy survivor. "The feeble go first," she says bluntly. Stewart feels that we must review previous conclusions. As she sees it, "the nub of the problem" is: "Can the straightforward analysis of dose-related death rates—which are the basis of current beliefs about the health risks of radiation workers—be regarded as robust under the conditions that existed . . . . The present [1982] review has found even the firmest of earlier conclusions unacceptable." And so Stewart has presented us with the simple, crucial variable of "selective survival of exceptionally fit individuals during the period of acute mortality after the A-bomb radiation."

In a 1977 report to the Nuclear Regulatory Commission, Dr. John Gofman cited "the possible existence of especially susceptible individuals" as "among the major factors that ultimately determine the amount of cancer arising from exposures of human populations to radiation."

In 1982, Dr. Jacob Fabrikant, another eminent radiation researcher, pointed out: "Calculations of . . . radiation induced cancer must take into account additional confounding factors, including sensitive genetic subgroups and exposure to other potentially carcinogenic agents." He noted that "host factors, environmental factors and immunologic factors also influence risk of cancer induction by radiation."

Dr. Irwin Bross wrote a letter to *The New York Times* hoping to publicize the shocking awareness a few researchers had arrived at. His own cancer research had enabled him to identify children who were up to twenty-five times more susceptible to leukemia (compared to the general population) as a result of x-ray exposure *(New England Journal of Medicine,* 287:107-110, July 1972). He stated: "Procedures for calculating 'safe levels' based on 'average exposure' of 'average individuals' are not going to protect the children or adults who need the protection most."

Whether you state the fact positively or negatively, it still amounts to the same thing—those who are healthy are the most fortified against radiation, those who are less healthy are more susceptible. But what does nutrition have to do with this?

Nutrition, Optimal Health and Protection Against Radiation

Good nutrition is directly related to good health. On the most basic level, our diets provide us with the energy (calories) and nutrients we need to carry on vital cellular processes. In the prolonged absence of adequate food intake, vitamin deficiency diseases, malnutrition, and starvation result. But diet can also enhance or impair our health in more subtle ways. For example, the quality of the food we eat can have a bearing on the efficiency of the body's immunological defense systems. In particular, nutrition can alter our susceptibility to various types of cancer and help us combat the proliferation of harmful free radicals.

Charles Simone, MD, author of the 1983 book *Cancer and Nutrition,* states: "Poor populations around the world who are malnourished are more susceptible to infection than those who receive adequate nutrition. Investigators studying the relationship between the immune system and nutrition have found that nutrition affects immunity and also affects the development of cancer either directly or indirectly via the immune system."

"Dietary Carcinogens and Anticarcinogens," an article published in *Science* on September 9, 1983, discusses factors in cancer incidence. Author Bruce Ames concludes that "epidemiologic studies have indicated that dietary practices are the most promising area to explore." Reviewing recent studies, including the National Academy of Sciences report *Diet, Nutrition, and Cancer* (1982), he notes: "These studies suggest that a general increase in consumption of fiber-rich cereals, vegetables, and fruits and decrease in consumption of fat-rich products and excessive alcohol would be prudent." Moreover, he points out: "we have many defense mechanisms to protect ourselves against mutagens and carcinogens, including continuous shedding of the surface layer of our skin, stomach, cornea, intestines and colon."

Good health is actually a state of continuous detoxification; our bodies maintain a constant defense against mutagens (substances that can cause chromosomes to mutate) and carcinogens (cancer-inducing agents). The healthy immune system destroys toxic substances that do get into the body, and isolates and removes damaged cells before malignancies develop. (We will examine the body's defense systems in more detail in Chapter 4.)

In a study entitled "Systemic Protection Against Radiation," published in *Radiation Research* in 1975, investigators observed radiation effects on two groups of rats, one fed an average diet and the other a high-nutrient diet. The latter group was found to be more radiation-resistant: "The radioprotective action of the diet might be explained in terms of an enhanced cellular proliferation of the hematopoietic tissues [tissues responsible for manufacturing new blood cells] . . . and a more efficient immune response in diet-fed animals." "However," the authors point out, "it is important to note that the diet must be fed for a certain time to achieve a radioprotective action."

Among the many researchers who have considered the interrelationship between nutrition and environmental hazards is Dr. Roger Williams, a pioneer in preventive medicine. Williams based his book *Nutrition Against Disease* on the thesis that the nutritional microenvironment of our body cells is crucially important to our health. He writes: "Those who are well nourished in every respect are likely to be protected against drugs and toxins that are foreign to their bodies." Dr. Carl Pfeiffer, an eminent scientist, also notes: "Researchers are discovering that dietary methods can help protect our bodies from environmental abuse."

And the well-known health reporter Herbert Bailey finds that "the natural defenses of living things against deleterious free radical reactions can be significantly influenced by diet." While researchers and scholars in many disciplines have found that if a person is nutritionally well-fortified he is less likely to absorb or be affected by dangerous substances, this awareness needs to become more widespread.

Healthy Survivors and Human Survival

George Ohsawa, a student of philosophy who contracted tuberculosis at the age of sixteen and was declared incurable by medical doctors, took it upon himself to discover what was wrong with him, and eventually found his way to health and then devoted his life to sharing his philosophy through writing and lecturing. In one of his books, *Cancer and the Philosophy of the Far East,* he says: "An organism with a stable constitution will be able to absorb and neutralize any unbalanced product—up to a point. In fact, we can even extend this theory to its ultimate by saying that healthy organisms can absorb and neutralize any poison. The historic case of the monk Rasputin is an example of this capacity pushed to the extreme."

Dr. Tatsuichiro Akizuki, author of *Nagasaki 1945,* the first full-length eyewitness account of the A-bomb attack on August 9, 1945, confirms the healthy survivor theory in his observations of those who came to him for help afterward: "All the people who fell victim to the bomb and who happened to be within 500 meters of the epicenter died before August 15. After that, nearly all those people who had been within 500 and 1,500 meters of the explosion slowly died, one after the other, between then and the end of September. The difference in their time of dying depended on the quantity of radioactive rays they absorbed, on their physical condition and their exact location at the moment of the explosion." He points out: "Their survival time depended on their constitution and age which all made a difference to the way in which radiation sickness attacked them and to the length of time they survived." This doctor and his staff survived and continued to care for others on a simple diet of rice and miso (a traditional food made from fermented soybeans and sea salt, often with the addition of grains such as barley or rice).

The *Radioecological Assessment of the Wyhl Nuclear Power Plant* (commonly referred to as the *Heidelberg Report),* was published in Germany in 1979. The report, which studied all aspects of nuclear plant emissions, states that the risk factor for radiation depends on many things, but as far as the individual goes the age, sex, state of health and genetic constitution are determinants. These researchers wrote: "Since not all individuals are healthy, some individuals may be exposed to significantly higher radiation doses than those calculated here for a normal, healthy person; this will depend on a person's genetic constitution and state of health."

Michio Kushi, a leading nutrition teacher and philosopher, feels that the future survival of the human race is threatened not so much by radiation and environmental hazards as it is by poor health induced by poor eating habits. He says: "Radiation and pollution can accelerate the accumulation and spread of cancer, but only if the body is already in a weakened state. The reason that some people would be more susceptible to the effects of radiation, for instance, is that the overall condition of their blood and tissues is not healthy as a result of their longtime dietary habits." He is realistic yet hopeful about human survival: "People still eating the modern refined diet have a much lower tolerance for radioactivity and are at risk of developing leukemia and other cancers. Reversing the biological degeneration of modern society is the key to curing atomic sickness and other forms of cancer."

4

Your Body's Radioprotective Systems

To enthusiastically implement a radioprotective way of eating it is helpful to understand the value of the body's physiological defenses against radiation. In this chapter we will examine the following aspects of the body's defenses: the purpose and function of the kidneys and liver, the significance of the immune system, and how the immune system is affected by various factors. In addition, we will consider the role that the acid-alkaline balance of the blood plays in protecting us from radiation.

The kidneys and liver are the two organs that are especially important to keep in good functioning condition because they detoxify poisons; if they are overworked and debilitated they cannot resist radiation. One way to take care of the body's radioprotective systems is to reduce the already existing load of toxins. This can be done gradually by following the recommendations given in this chapter and the eating plan outlined in Chapter 6. It is equally important to maintain a constant state of detoxification by assiduously keeping up an intake of the suggested foods.

Taking The Burden Off Your Kidneys

Among the body's avenues of elimination, which include the intestinal tract, lungs and skin, the kidneys have a particular radioprotective value. The kidneys are a remarkable filter system that serves to cleanse the blood of toxins and waste. They also produce urine to facilitate the elimination of these poisons from the body. In addition, the kidneys regulate the overall composition of body fluids, including the acid-alkaline (pH) balance of the blood. This pH balance affects our susceptibility to radiation.

About forty-five gallons of liquid daily go through these small organs, which are about the size of your clenched fist, and about one and a half quarts are excreted. Toxic substances such as mercury, which is often found in tuna and other large fish, can destroy some kidney cells. Other substances that can damage the kidney include arsenic, oxalic acid, cadmium and lead.

A 1982 article on cadmium in *Archives of Environmental Health* tells us that "with chronic low level exposure the kidney is considered to be the organ at risk." Two years later, the same journal published a study with the finding that 3.9 million Americans had possible occupational exposure to substances known to poison the kidneys—particularly uranium, silica and organic solvents, in addition to lead, cadmium and mercury.

Stress and diets high in meat are two other factors that can also overwork the kidneys. The kidneys cannot work properly when there is too much acid in the body. This can result from stress or from excess reabsorption of toxic acids from the bowel. Whatever the cause, the fact remains that when the kidneys do not function well, the other eliminative organs work overtime.

Too much liquid overwhelms the delicate filtering work of the kidneys. The often-heard recommendation to "drink eight glasses of water a day" is most appropriate for a person who consumes a diet rich in meat. Such an individual needs that much liquid just to flush out the toxic acids that are the byproducts of protein and fat metabolism. It is a mistake to drink too much liquid to combat this condition, as this floods the kidneys and inhibits their filtering work. It is better to remove the cause by reducing the intake of meat. When we rely on grains, vegetables and beans as our primary sources of protein, extremes such as eight glasses of water a day are not necessary—it is much more beneficial to drink less and allow the kidneys to operate at their best potential.

If waste cannot be excreted through urine it is often eliminated through the skin by perspiration. For this reason the skin is often referred to as the "third kidney."

Nutrients that improve kidney functioning are calcium, vitamin B$_6$, vitamin C, and magnesium in particular. Buckwheat is one food that can supply the kidney with several of the nutrients it needs.

Taking The Burden Off Your Liver

The liver does an astounding amount and variety of work. Although it has over 500 known functions, we will focus on just two of them here. It absorbs and sorts out nutrients such as fats and sugars, and neutralizes toxins. It cannot filter out radiation, but due to its work on nutrients and toxins, the liver has a major effect on the quality of the blood.

Diet has a wide-ranging influence on liver function. For example, the liver secretes bile, which emulsifies fats and breaks them down. However, excess fats are stored in the liver. The liver (along with the pancreas and other organs) also helps to regulate the amount of sugar (glucose) in the blood. It stores excess sugar in the form of glycogen. If the diet is high in refined carbohydrates (including sugar), the liver's capacity is strained. When this happens the liver releases sugar into the bloodstream.

A diet low in sugars and fats and with a minimum of chemical food additives is an aid to the good functioning of the liver. Overeating and overuse of alcohol are an extra strain. Eating too much meat, eggs and cheese can damage the liver; so can eating too much fruit and refined sugar. Each extreme can cause a different type of malfunction. A balance of more neutral foods will come to the rescue. The fewer toxins in the diet the better, because the liver also has to handle the all-pervasive contaminants that we can't avoid—including radiation.

If the liver is overworked by the need to neutralize many toxins and at the same time undersupplied with nutrients that it needs to function, all of its vital operations can be impaired. When this happens, poisons can accumulate in the liver and consequently are not sent out to be filtered and eliminated via the kidneys. The liver can become hard and swollen. Poisons that remain in the body lower our overall level of health and thus make us more vulnerable to the effects of radiation. The healthier the liver is, the healthier we are overall. Thus, the condition of the liver is important in radiation protection, especially in terms of the healthy survivor effect discussed in Chapter 3. One of my naturopathic teachers used to say "You live in your liver."

Physical signs of liver damage are yellow in the eyes and sometimes a large crease, usually thought of as a smile line, down the middle of the cheek. Traditionally, an angry temperament has also been regarded as a sign of liver damage (the dual meaning of the word *bilious* reflects this perception).

Nutrients especially needed for optimal functioning of the liver are all the B vitamins, vitamin C, and methionine, a sulfur-containing amino acid. The latter two, both used in detoxification, are found in some vegetables (see Chapter 6).

Ways to help a tired liver include eating daikon radishes (long white radishes), and dark green leafy vegetables daily. It is also helpful to avoid meat, eggs, cheese, chocolate and fatty foods in general, along with sugars, alcohol and foods that have been sprayed with pesticides. Increased use of the foods highlighted in Chapter 6, especially miso and tempeh, can also benefit the liver.

The Immune System:
The Front Line Of Defense

The immune system has received worldwide attention lately because of the AIDS epidemic. The devastating effects of AIDS, however, are just a part of a larger picture. It is of the utmost importance for us to maintain the immune system at its highest functioning level, so that we will be prepared at all times to withstand the constant assaults of low-level radiation.

The immune system is a complex regulatory system that protects us from viruses, bacteria and other microorganisms, allergens (substances that provoke allergic reactions), toxins, and the growth of malignant cells. It includes the spleen, thymus, bone marrow, blood, and lymph nodes. Some people are said to have *strong immunity.* This means that their immune system is functioning well and ever-present viruses and cancer cells are not allowed to get a foothold.

This protection is orchestrated by the thymus gland. The tissues of. the bone marrow form *lymphocytes* (a kind of white blood cell). Some lymphocytes are known as *B-cells*—the "B" stands for bone. Some of these pass through the thymus gland, where they mature into *T-cells* (the "T" is for thymus). Along with a few other types of immunoprotective blood cells, B and T cells circulate in the bloodstream. The B-cells produce *antibodies,* providing immunity against specific infectious agents; the T-cells kill foreign cells and cancer cells.

The *spleen* is an organ composed of lymphoid tissue. Its functions are to destroy old or defective red blood cells and to form some mature lymphocytes. Some lymphocytes also mature in the *lymph nodes,* but the main function of these areas of the body is to act as filters. The lymph nodes catch foreign particles and bacteria and keep them from circulating around the body. The tonsils are lymph nodes. If you've ever had inflamed tonsils, it was a sign that the immune system was overworked. The best solution is not to remove the tonsils, but to eliminate the cause of the congestion. Other lymph nodes are located in the armpits, groin, neck and spleen, and along the digestive tract.

Distinguishing Self from Non-Self

How does all this work to protect you? The thymus, bone marrow, spleen and lymph nodes manage to distinguish *self from non-self.* That is, any foreign invader, be it a virus, cancer cell or toxic substance, is recognized as "not good for the body"—and is attacked, surrounded with antibody-producing B-cells, by T-cells, or by other specialized "defensive warriors," such as enzymes. Once the invader is destroyed, it is carried through the lymph and eliminated. In this way we protect our internal environment against problems from the external environment.

Radiation Exposure Reduces Immune System Function

Ionizing radiation can cause varying degrees of damage to the immune system. While a small dose may lower the functioning an imperceptible amount, it may nonetheless result in shortened lifespan, lowered resistance and diminished quality of life. In cumulative small amounts, radiation can reduce body functioning to such a level that cancer is allowed to take hold and many other ailments can get started.

Here is how this works. When radiation strikes the body, it causes damage to the cells. Radiation kills some cells outright. It also knocks electrons out of their atomic orbits. The loss of an electron can turn an atom into a free radical which is reactive with other atoms and molecules. This can lead to extensive cellular damage, for two reasons. First, each cell of the body contains vast numbers of atoms that can lose electrons. Second, the formation of a free radical starts a chain reaction leading to the proliferation of free radicals.

These free radicals can cause a number of problems (see pages 40-41). In terms of immunity, the biggest problem is the mis-synthesis of proteins. Proteins that contain mistakes are misconstrued by the immune system. Recognizing that such mis-synthesized proteins are not like other body proteins, the immune system perceives them as "non-self"—dangerous outside substances—and sets to work to clean them up and get rid of them. If there are not too many free radicals it can manage. If there are too many, the system becomes overloaded and it becomes unable to clean up actual foreign invaders. When this happens, viruses, microbes, cancer cells and any kind of infection can take hold.

The Thymus

The thymus is a small gland in the chest located behind the breastbone, just above where you think your heart might be. Until recently it was thought to be a gland of childhood that didn't serve any function after puberty. At about the age of ten, the thymus stops growing and starts to atrophy. Nevertheless, in the 1960s researchers began to discover that the thymus has a continuing input throughout adult life. It is, in fact, the key to a strong immunity.

As described in a 1968 issue of the journal *Cancer*, doctors studying the relation between tumors in the thymus and the development of cancer found: "*A* relationship between the thymus gland and malignant disease has been suggested by numerous experimental studies, and abnormalities of the thymus have been related to a variety of . . . diseases." They concluded: "The association raises the possibility that the thymus plays a role in the pathogenesis [origin and development] of malignant disease in man as well as experimental animals." Another study in *Cancer*, published the following year, offered a similar conclusion: "The findings suggest there is a relationship between thymus and cancer in animals and man in which the thymus has a pathogenetic importance."

The researcher Jean Marx has proposed that the thymus is the "biological clock" that determines how fast we age. His article "Aging Research: Pacemakers for Aging?" was published in a 1974 issue of *Science*. In a 1975 article, also appearing in *Science,* Marx noted: "The thymus is necessary for the normal differentiation and maturation of T cells." As we know, T-cells are indispensable for strong immunity and internal balance. Marx feels that the thymus hormone given to cancer patients receiving radiation treatments could prevent the resulting decline of T-cells. This seems logical, but for those of us in general good health, it is better to obtain plenty of the nutrients that reinforce T-cell function—vitamin B_6 and zinc.

How To Maintain Your Immune System

There are some variables that can reduce your immunity. In addition to kidney and liver damage, these are: deficiencies of zinc and vitamin B_6; a poor diet overall; a high-fat diet; inferior drinking water; stress; and the deleterious effects of radiation and pollution. The best ways to maintain your immune system is to be conscious of these variables, to make every effort to maintain the optimal level of nutrition and health, and to minimize your exposure to environmental hazards.

Get Enough Vitamin B_6 and Zinc

Two nutrients that are absolutely necessary for the thymus are vitamin B_6 and zinc. For example, zinc-deficient children have been found to have atrophy of the thymus gland. A 1977 study of undernourished children who were given zinc supplements showed an increase in the size of the thymus. This study, which was published in the *Lancet,* states: "Atrophy of the thymus has been associated with zinc deficiency, pyridoxine (B_6) deficiency and a wide variety of acute and chronic illnesses." (Vitamin B_6 and zinc will be discussed more fully in Chapter 7.)

Overall Diet

Beside the two particular nutrients, zinc and vitamin B_6, the diet in general is now known to have an effect on immunity. "Diet and Immunity," one of the most important studies on this subject, appeared in the *American Journal of Clinical Nutrition* in 1979, and it states: "The relationship between nutrition and immunology has only recently received attention, but, as the results of this study emphasize, the relationship is a crucial one." A 1981 study in *Nutrition Reviews* says: "On a global scale under nutrition is the most common cause of immune deficiency. The knowledge that immunity is compromised in states of nutritional deprivation is relatively recent." The Diet for the Atomic Age offers optimal nutrition and immune support.

Reduce Fat Intake

Throughout the industrialized world, it is generally acknowledged these days that it is advisable for the average adult to reduce his or her intake of fats. Usually this is recommended in reference to maintaining healthy arteries and avoiding heart trouble. However, recent research indicates that an excessive dietary intake of lipids (fats) can impair the immune system. "Lipids and Immune Function," a study published in *Cancer Research,* finds that fats reduce immune function. According to the authors, "This offers the promise that by changing dietary lipids one may be able to modulate immune function and thus alter an individual's susceptibility or resistance to diseases." As fats have over twice as many calories as either proteins or carbohydrates, they are a prime contributor to obesity. This in itself can be a problem—at least one study on obese animals has shown they have a reduced immunity. The National Academy of Sciences put "Reduce the intake of dietary fat" as *number one* on the list of six important dietary recommendations for reducing cancer risk.

Find Your Way to Less Stress

In addition to nutrients (but less easy to control), our *attitudes* also exert an influence on the immune system. If we are worried and under stress our bodies react; the efficiency of the immune system declines. In his popular book *Anatomy of An Illness,* Norman Cousins told how he utilized this fact to help overcome what was thought to be a terminal illness—by moving out of the stress of the hospital environment and into a hotel, where he was able to relax and cheer himself up by watching Marx Brothers movies.

Stress causes biochemical changes in the body. For example, the adrenal gland produces hormones called corticosteroids in response to anxiety. One of the effects of these hormones is to decrease the body's production of lymphocytes, which mature in the thymus, spleen and lymph nodes. A June 1981 article in *Science* notes: "The primary harm that corticoids cause when in high concentration is damage to lymphocytes and thymus elements that are essential for optimum cell mediated immune defenses."

Another biochemical effect of stress is increased acidity in the internal environment. If such excess acidity persists for an extended period of time, it can create a number of problems, including impaired kidney function. An acid body chemistry can also predispose us to radioactive uptake.

Avoid Exposure to Environmental Contaminants

Impaired immune function results from exposure to chemicals, and may be a delayed effect of early exposure. The Heidelberg Report has this to say about the relation of pollution to radiation resistance:

> Population groups in which previous damage has been caused by other noxae [noxious substances; i.e., injurious to health] must . . . be regarded as critical groups. This means that the effect of the additional radiation exposure will be greater in these groups than in groups in which there is no pre-existing damage.

In terms of radioprotection, the fundamental problem is "that pollutants damage our cleansing organs (the liver and kidneys) and tax the immune system, so that our defenses are down. We are then all the more vulnerable to radiation. Clearly, we must reduce general pollution as much as we can. There will inevitably be some radioactive pollution that we cannot escape, and we want to be best able to keep it from making its inroads.

On a day-to-day basis, the best way to minimize our exposure to environmental hazards is to be aware of how they affect our individual lifestyles and to make any adjustments accordingly—for example (depending on where you live), by buying locally grown produce or using bottled spring water for drinking. We can also make sure we use solvents and other household chemicals only when necessary and always in a well-ventilated area.

There are so many environmental hazards besetting us on all sides that it would take an entire book to discuss them all! Focusing on the body's radioprotective systems, however, it is important to examine two things: the effects of toxic metals and the quality of our water supply.

Pollution And Its Effects

When I was a kid, in the summer mosquito season at the Jersey shore, a truck often came lumbering down the road just behind our beach house, letting out a wide fog of DDT that drifted over the bayberry bushes and swampy grass areas where the mosquitoes bred. The smell bothered us temporarily, but we were glad to get fewer insect bites.

The pollution of our planet has escalated in the most extraordinary way since those summers. New technology in industry and agriculture is churning out poisons that contaminate our environment at an unprecedented rate. We have learned a few lessons—as the 1974 banning of DDT shows—but our croplands are still besieged by synthetically compounded pesticides, herbicides, fungicides, fumigants, fertilizers and other chemicals. Our air and water are still being filled with toxic fumes, chemicals and radioactive wastes that are byproducts of many types of industry. Even automobile exhausts are taking a toll on our surroundings.

Since World War II, production of chemicals in the United States has increased from 2 billion to 360 billion pounds per year. Modern farming is destroying the soil: about a billion pounds of pesticides are used in this country every year. Many experts feel that most of the 3,400 pesticides now in use have not been adequately tested.

Pesticide residues become concentrated along the food chain (see pages 50-51) and turn up everywhere, including in mother's milk!

In the spring of 1984, there was a flurry of concern about residues of the chemical ethylene dibromide (EDB), a fumigant that is widely used on grains. Well-known brands of cake mix, flour and other products were whisked off supermarket shelves. But it was too late to reach wheat flour that had been shipped all over the country as part of Federal food subsidies, some of it going to school lunch programs.

The Environmental Protection Agency (EPA) has a list of 22,000 suspected hazardous waste sites, and the General Accounting Office estimates that 378,000 toxic sites will ultimately be discovered. The process of discovering these sites is an ongoing one, as dumping of hazardous wastes has gone on for many years and the culprits are reluctant to take responsibility. Nevertheless, once identified, these sites need to be cleaned up. Despite the Superfund program, progress here has been very slow. Of 849 recognized hazardous waste dumps slated for cleaning up, only 10 have been cleaned and closed. Environmentalists contend that the cleanups have been less than adequate.

In March 1986, 511 land disposal facilities for hazardous wastes were seeking official permits from the EPA. In the meantime, they were operating on "interim status" permits. In July 1984, a report entitled *America's Toxic Protection Gap* (which was endorsed by several former government officials and about thirty organizations such as the PTA, environmental groups and religious groups) revealed that 75 percent of such "interim" facilities had violated the minimum safety standards. It seems that little improvement has been made since that time.

Our land disposal facilities for hazardous wastes are overflowing. According to the EPA, 1,138 sites were closed or closing as of March 1986. Over 28,000 tons of garbage daily are tossed into landfills in New York City alone. The only solution is to build "solid-waste reclamation centers"—incinerators which emit poisonous fumes.

The quality of the air we breathe is yet another cause for concern. In recent years, about 80 percent of the industrial plants tested have been found to have emissions above the standards allowed for air pollution. With all this in mind, let's not talk too much about food additives, artificial coloring, chemical preservatives and all that!

Toxic Metals

There are minerals that are essential to life, such as calcium and magnesium. There are minerals essential in a small amount, but toxic in a larger amount, such as copper. And there are elements that have no nutritional value and are toxic in any amount. This group includes lead, cadmium, mercury and aluminum.

These toxic metals are a constant impediment to maintaining a high level of health. Although there is extensive medical research to support their harm, they are not usually thought of in a diagnosis by a physician. After counseling nutrition clients for many years, I have come to believe that possible metal toxic-ity is the first thing to rule out.

Also, most of the voices for health and nutrition that we hear today speak about everything *but* the level of toxic metals that might be in the body, reducing the functioning level of the organs. No matter what amino acid may improve brain function, no matter what nutrient may help to increase muscle mass and reduce fat, we will still be operating on a lowered potential as long as we have weakened immune systems. Our liver and kidneys need to be operative so they can work to constantly maintain a detoxified body. The amount of radiation in our environment makes it imperative that we cleanse ourselves—and cleanse ourselves constantly.

Lead

A number of experts believe that lead was a factor contributing to the fall of the Greek and Roman empires. In classical times, water was collected from lead-covered roofs by lead gutters that ran into lead-lined containers. Grapes were boiled in lead pots in the making of wine. Many salves, ointments and cosmetics and some paints contained lead.

All this may have caused mental decline and lowered birth rates among the ruling classes. Now some scientists believe that our modern civilization is going the way of ancient Greece and Rome. Epidemiologists, who study the distribution of diseases in large groups of people, generally agree that lead is the worst pollutant element. It is one of the most widely produced metals. However, production figures tell only part of the story; about half of the lead used today is recycled. Lead consumption in the United States reached a peak of over 1.4 million metric tons per year in 1978 and 1979. The 1980 book *Lead in the Environment,* prepared by the National Academy of Sciences, predicted:

> It will be shown in the future that the average American adult experiences a variety of significant physiological and intellectual dysfunctions caused by long-term chronic lead insult to their bodies and minds which results from excess exposures to industrial lead that are five-hundred fold above the natural levels of lead exposure and that such dysfunctions on this massive scale have significantly influenced the course of American history.

It has been estimated that about 38 million Americans have a significant amount of lead in their bodies. Most large metropolitan areas have established screening programs, as children are most vulnerable to lead toxicity. Lead poisoning can affect the infant *in utero* and disturb the growth and behavior of newborns.

A nationwide four-year study published in the *New England Journal of Medicine* in 1983 reported that 4 percent of all children aged one to six, 11 percent of all inner-city children and 18 percent of low-income urban Black children had undetected lead toxicity. "Undetected" lead toxicity means that no overt symptoms are expressed, but subtle effects of poisoning are noticeable. Since that study appeared, these figures have changed considerably as a result of two trends. On the bright side, the amount of lead in gasoline was reduced by nearly 95 percent between 1975 and 1986. But the threshold level for health damage due to lead has been revised downward. At the time of the above-mentioned study, levels of 30 micrograms of lead per deciliter of blood were regarded as cause for concern. Today, the recognized level is 25 mcg/dl.

Like radioactive substances, lead is a cumulative poison. Once it is taken up by the body, lead accumulates in the bones, liver

Lead: Sources, Effects and Protective Factors

Source	Potential Effects	Protective Factors
airplane exhaust	abdominal cramps	B-complex vitamins
artist's paints	anemia	vitamin C
automobile batteries	arthritis	vitamin D
bone meal	brain damage	calcium
pills ceramic	hyperactivity	magnesium
glazes on pottery	hypertension	zinc
cigarette smoke	interference with	adequate fiber
dust and chips from	formation of	(especially
lead paint	hemoglobin	pectin)
insecticides	impairment of female	sodium alginate
lead pipes lead	reproductive	cabbage family
smelting	capacity	vegetables
leaded gasoline	impairment of growth	
(automobile	and behavior of	
exhaust)	newborns	
solder	interference with	
vegetables from	body's synthesis	
roadside gardens	of vitamin	
	D, leading	
	to calcium	
	deficiency	
	kidney damage	
	liver damage	
	loss of appetite	
	neurological damage	
	palsy, paralysis	
	psychological	
	problems,	
	mental disturbances	
	reduction of	
	immune function	
	weakness	

and kidneys. Even moderate levels can cause kidney damage and suppress immune function. Overt symptoms of lead poisoning include severe weakness, abdominal cramps and paralysis. Less overt but equally debilitating are the effects lead has on the bloodstream. It interferes with the formation of hemoglobin and causes anemia. It also causes mental imbalances.

The U.S. Government and American industry are at an impasse regarding lead in the environment. Though our lead consumption has remained fairly constant at about 1.1 million metric tons through the 1980s, a gradual increase is forecast. One reason for this projected increase is that technology has not yet found an alternative to the lead-acid batteries used in motor vehicles. Safe disposal of these batteries has become a major problem.

It is important for us to be aware of the paradox presented by our society's need for lead and the health hazards lead causes. We can insist that solutions actively be sought. We can also protect ourselves by eliminating problems in our home environment.

Today, chips and dust from lead-based house paints are the major sources of acute lead poisoning. Drinking water supplied through leaded pipes is also a potential source of acute exposure. It is wise to avoid lead-based paints and water from leaded pipes as much as possible. Gasoline remains a background source of lead in the environment. Other sources are listed in the accompanying table, along with some potential effects of excessive exposure and some protective nutrients.

Mercury

Starting in the days of the Roman Empire and continuing into the present, thousands of uses have been found for mercury. In ancient Rome, mercury was mined for use in separating precious silver and gold from other ores. From ancient times until the recent past, workers in mines and refineries suffered one way or another. Other occupations suffered as well. The phrase "mad as a hatter" reflects the fact that hatters, who were constantly exposed to mercury compounds used in making felt hats, often went insane.

In addition to insanity, symptoms of mercury poisoning include blindness, loss of teeth, brain damage, loss of motor control, and, with consistent exposure, possible coma and death.

The United States has used more than 160 million pounds of mercury since 1900 in chemical, agricultural and industrial areas. Mercury is a cumulative poison. It gets into the environment in the form of toxic vapors or in the deadly organic form

Mercury: Sources, Effects and Protective Factors

Source	Potential Effects	Protective Factors
chemical fertilizers	allergies of all kinds	selenium
contaminated tuna fish, swordfish, and other large fish	arthritis	adequate fiber
	birth defects	good overall nutrition
	brain damage	cabbage family vegetables
dental amalgams (fillings)	collagen problems with elbows and knees constriction of vision, cataracts, blindness	
explosives		
floor waxes		
fungicides		
industry	depression	
ointments and some cosmetics (especially skin-lightening creams)	harm to developing fetus	
	kidney damage	
	loosening of teeth	
pesticides	neurological damage resulting in epilepsy, seizures, multiple sclerosis	
Pharmaceuticals		
photographic film		
plastics	reduction of immune function	
water-based paints	reduction of white blood cell count	
	weight gain	

known as methyl mercury. Methyl mercury goes into the water and accumulates in the food chain. It has been known to contaminate swordfish and other large fish at the top of the food chain. Thus, I feel that the tuna in sushi is best left aside and tuna salad sandwiches are best avoided. The sources and potential effects of mercury exposure are summarized in the table above.

Some mercury goes directly into our internal environment in the form of "silver" dental fillings. Some silver amalgam fillings are 40 to 50 percent mercury. This is being absorbed continually—a bit like having a poison implant! Fortunately, other materials are now in use for dental fillings.

The overall effect of mercury on the immune system is to lower it. It depresses the number of white blood cells, including T-cells, which kill foreign invaders. In a May 1984 article in the *Journal of Prosthetic Dentistry,* Dr. David Eggleston reported that when six silver amalgam fillings were removed from a patient, a 26 percent increase in the number of T-cells resulted. Just to be sure this wasn't a coincidence, Eggleston put four silver fillings back in. The result? An 18 percent decrease in the T-cell count.

Cadmium

Cadmium is a tin-like metal that has become a threat to our nation's health. It may be an even greater threat than lead. Cadmium dust and vapors are highly toxic. Exposure leads to serious kidney and lung damage and has been linked to cancer. The National Institute for Occupational Safety and Health has estimated that about 1.5 million workers are currently exposed to cadmium.

About 700 tons of cadmium are emitted into the air each year from incinerators, smelters, and other industrial processes, according to recent estimates by the Environmental Protection Agency. The metal is used in a variety of ways: in nickel-cadmium batteries, in solder and other alloys, and as a pigment

Cadmium: Sources, Effects and Protective Factors

Source	Potential Effects	Protective Factors
cigar smoke	antibody suppression	vitamin C and other
cigarette smoke	dry skin	detoxifiers
drinking water	emphysema	calcium
fertilizers	heart problems	selenium
garden soil	high blood pressure	zinc
	(hypertension)	

incinerators	impaired calcium metabolism	adequate fiber
industrial air pollution	kidney damage	cabbage family vegetables
metallurgy	loss of hair	
pipe smoke	loss of zinc	
refined grains		

---

and stabilizer in plastics. It is also electroplated onto steel to prevent corrosion.

Several recent studies have demonstrated that cigarette smokers have significantly high cadmium levels in their bodies. "Passive smokers"—such as a smoker's family and co-workers—also accumulate cadmium. Cadmium is a cumulative poison and can result in varying degrees of toxicity. It induces hypertension (high blood pressure), accumulates in the kidneys and suppresses immunity. Other manifestations of excessive cadmium exposure are shortened lifespan, anemia, interference with zinc metabolism, decreased T-cell production and liver damage. Overall immunity is decreased due to the damage to the key organs in the immune system (liver and kidneys) and the decline in T-cells. The table on page 95 lists some cadmium sources and potential health effects of cadmium exposure.

Aluminum

While not a heavy metal (like lead, mercury and cadmium), aluminum has recently been recognized as a toxic metal. Because for many years no one realized that it was absorbed by the body, aluminum has been, and still is, used to make pots, pans, cooking utensils and foil. Aluminum is also an ingredient in many over-the-counter medicines and is an additive (in the form of sodium aluminum sulfate) in most commercial brands of baking powder. It is sometimes present in drinking water. Elimination of aluminum sources such as cookware, aluminum foil, antacids and baking powder is important because there are unavoidable traces of naturally occurring aluminum in all foods. Like other contaminants, aluminum increases in concentration along the food chain. In patients with Alzheimer's disease (senile dementia), four times the normal amount of aluminum accumulates in nerve cells in the brain. Exposure to high levels of aluminum may contribute to the development of Alzheimer's. In addition, aluminum has been implicated in extreme hyperactivity and behavior problems in children, anemia, headaches, liver problems, kidney problems including dementia in dialysis patients, colitis, and the neurological changes associated with Parkinson's disease. The accompanying table summarizes common sources of aluminum exposure, potential effects and protective factors.

Aluminum: Sources, Effects and Protective Factors

Source	Potential Effects	Protective Factors
aluminum cans	anemia	vitamin C
aluminum foil	Alzheimer's disease	calcium
antacids	brain and nerve cell changes	zinc
baking powder	colitis	adequate fiber
with sodium	deactivation of parathyroid	
aluminum	gland	
sulfate	dialysis dementia	
buffered aspirin	headaches	
certain cheeses	hyperactivity in children	
cooking utensils		
deodorants	impaired thyroid function	
drinking water	juvenile delinquency	
junk foods	liver and kidney problems	
pots and pans	lower calcium levels	
regular table salt	neurological changes	

Avoid Tap Water for Drinking

One way in which pollution affects us daily is in the quality of our water supply. Although the Environmental Protection Agency, in the National Water Quality Inventory—1984 Report to Congress, pronounced "significant improvements in water quality," it also admitted that serious pollution blighted many rivers, lakes and reservoirs. Thirty-three states reported toxic substances in the water.

Our drinking water may contain more than 500 chemicals from agriculture and industry, some of which have been proven to cause cancer. About 60 percent of our water has added fluoride, which prevents tooth decay in children, but can ultimately damage the immune system. Unlike chlorine, an antibacterial additive directed at affecting the water, fluoride is directed at the people who drink the water—particularly, children and adolescents who have not yet formed all their permanent teeth.

Fluorides are certain compounds containing fluorine, which is the most reactive element. Fluoride compounds are emitted into the air in gaseous and particulate (particle) form as waste from aluminum processing and other industries. The toxicity of the various fluorides varies, but exposure to any of them can have negative health effects.

Fluorinated hydrocarbons, for example, can irritate the skin and be absorbed by certain organs. Another compound, sodium fluoride, was first used as a rat poison. But in the 1940s, it was promoted as an additive to the water supply that would reduce tooth decay. In spite of objections from the American Medical Association and the American Dental Association, it began to be added to the water (a process known as fluoridation). An editorial in the October 1, 1944 *Journal of the American Dental Association* noted that water containing from 1.2 to 3.0 parts per million of fluorine caused developmental problems in children. These effects included abnormal hardening and formation of the bones and teeth. The editorial stated:

> We cannot afford to run the risk of producing such serious systemic disturbances in applying what is at present a doubtful procedure intended to prevent development of dental disfigurement among children.

The book *Fluoride: The Aging Factor,* by Dr. John Yiamouyiannis, is must reading for anyone interested in how fluoride reduces immune function. Although years of research have established 1.5 parts per million (ppm) as a conservative "safety" level, Drs. Alfred and Nell Taylor at the University of Texas found that even smaller amounts (.5 to 1.0 ppm) increased tumor growth rate in laboratory mice by 15 to 25 percent.

Nevertheless, New York City alone now adds 55,000 pounds of fluoride to its water every day. Such fluoridated water usually contains about one part fluorine per one million parts water. While the American Dental Association no longer opposes fluoridation, there are indications that the line between the dose in our water and the harmful levels are fine indeed. The table on the next page lists a number of adverse effects that can be caused by exposure to fluoride from various sources.

Moreover, in the mid-1970s, a new concern arose—the possibility that fluoride might react with pollutants that had found

Fluoride: Sources, Effects and Protective Factors

Source	Potential Effects	Protective Factors
Beverages bottled in areas using fluoridated water	Abnormal hardening of the bones and teeth	Vitamin C
Drinking water	Acceleration of aging	Vitamin E
Fluoride treatments	Arthritis	
Industry	Birth defects	
Mouthwashers	Breakdown of collagen	
Some fertilizers	Damage to immune system	
toothpaste	Damage to thyroid gland	
	Destruction of cells	
	Destruction of enzymes	
	Docility	
	Genetic damage	
	Increase in torn ligaments	
	Increased risk of cancer	
	Kidney damage	
	Mental lethargy, lack of will	
	Mottling of tooth enamel	

their way into the drinking water supply, thus creating additional toxic compounds. Even chlorine is coming under scrutiny now, because it can complex with substances found in the water supply. For example, it complexes with plutonium and in so doing increases its toxicity up to one thousand times.

Fluoride is one reason to avoid drinking tap water. There are other reasons as well. For instance, tap water is often acidic. Acidic water tends to draw copper out of copper pipes (which are often used for water). Copper is an antagonist to zinc, a mineral that is necessary for immune function.

The best advice is to get a water purifier or bottled water. Before you do so, you may want to investigate the quality of your local water supply. Consider the possible presence of toxic chemicals, radioactivity, sewage and bacteria; find out if the water flows through lead or copper pipes; ascertain its mineral content and acidity. You may be able to obtain information from an environmental group or organization in your area. Whether you use a water purifier or bottled water, be an informed consumer. Preferably, bottled water should come from the deepest springs—ones that draw water from far below the level of the underground water table. Although such springs can be contaminated with nitrates, in general they are least apt to be polluted. It is important that the source of your drinking water be as uncontaminated as possible.

Aids And The Immune System

In the spring of 1981, epidemiologists recognized a new disease and named it Acquired Immunodeficiency Syndrome. AIDS, as it is commonly known, is a breakdown of the immune system that is associated with systemic infections, infections of the central nervous system and unusual types of cancer. People with AIDS are unable to fight off infections—major or minor. Their T-cells, B-cells and other immune defenses are suppressed and the immune system loses the ability to distinguish between self and non-self. The syndrome has a mortality rate that approaches 70 percent.

In October 1986 there were more than 27,000 reported cases of AIDS in the United States; since 1981, 11,000 people in this country have died as a result of acquired immunodeficiency. During the past few years, the incidence of AIDS has doubled as frequently as every six months. Despite some signs that the number of AIDS cases is leveling off, most authorities predict that there will be a tenfold increase in AIDS cases and deaths from the disease in the next five years.

Besides those with actual cases of AIDS, there are millions of AIDS carriers and a group of people with a condition called pre-AIDS, who have the potential to communicate the disease to many others unknowingly. AIDS has a long incubation period. Years may elapse before a person who has been infected with AIDS shows any signs of disease. And not everyone who is exposed to AIDS comes down with it.

What goes wrong in AIDS? There has been an extraordinary amount of speculation regarding the causes of this devastating and frightening disease. As a 1983 article in the *New England Journal of Medicine* observed: "tension has increased, as has the sense of mystery and confusion."

In addition, there is a great sense of competition among experts. It seems to me that the central question can be stated in simple terms. Does AIDS cause a weakened immune system—or does a weakened immune system cause AIDS?

Most researchers on the mystery of AIDS say that the virus HTLV-III causes persistent clinical immune deficiency—but conversely, might it not be that a chronic immune deficiency causes the virus to take hold? Drs. Frank Rosner and Jose Giron pointed to this possibility. Their writing appeared in the December 1983 issue of the *Journal of the American Medical Association:*

> As with many other diseases, the etiology [cause of AIDS] may turn out to be multifactoral, i.e., a genetic predisposition (?HLA type) and/or otherwise susceptible host (immunodeficiency) who is exposed to an environmental agent (?virus) with or without a triggering mechanism (?another virus).

The "genetic predisposition" which the authors speculate might play a role is the type of HLA. These letters stand for *human leukocyte antigen.* Antigens are substances that stimulate the B-cells' production of antibodies. There are different types of HLA. They are genetically determined and characteristic of an individual, like blood types and fingerprints. Some individuals may have a type of HLA that predisposes them to AIDS.

Many doctors working with AIDS patients have observed that the AIDS virus will produce AIDS only if the immune system is already damaged. One such physician, Dr. Joseph Sonnabend, one of the discoverers of interferon and a director of the AIDS Medical Foundation in New York City, feels that AIDS does have multifactoral origins. In "The Plague Years," an article in the April 1985 *Rolling Stone*, Sonnabend told reporter David Black:

> Since I've been watching the disease in this city, it never occurred to me that this disease could be a specific syndrome, a new infectious agent. The patients getting sick had been exposed to an extremely complicated biological environment.

One of the complicating factors in the worldwide biological environment is radiation. In November 1985, Dr. Ernest Sternglass addressed this point at the Brussels International Symposium on AIDS. Sternglass proposed that uterine exposure to radioactive fallout from atmospheric bomb testing in the 1950s and 1960s might be a predisposing factor in the incidence of AIDS in the 1980s. In other words, Sternglass argued that reduced immunity (which is known to result when a developing fetus is exposed to radiation) could be a factor that, when combined with further weakening influences, would set the stage for what is now labeled AIDS.

As we have seen, when radioactive strontium-90 is absorbed, it goes to the bone marrow and inhibits the healthy production of the protective white blood cells that are fundamental to a strong immune system. The level of strontium in the bones is affected by individuals' early diets (and by those of their mothers during pregnancy). Those who have adequate calcium are less likely to take up strontium.

I spoke with Dr. Sternglass in July 1986, and he explained the significance of new data concerning the AIDS epidemic in the Congo in Africa. This area, which is believed to be where the AIDS epidemic began, received heavy radioactive fallout (brought down from the upper atmosphere by rainfall) after atmospheric weapons testing in the late 1950s. In fact, while radioactive strontium-90 from bomb testing has been deposited in the bones of people all over the world, the highest ratio of strontium-90 to calcium is found in the bones of people in the Congo. Another factor that might weaken the immunity of people in the Congo is background radiation from the high-grade uranium deposits that have been mined in that region.

AIDS appears to be occurring to a great extent among the generation born from 1951 to 1963. Perhaps the dramatic increase in AIDS is because those individuals—who were exposed to the highest levels of radioactive fallout before they were born—have become sexually mature, have been exposed to sexually transmitted diseases, and have therefore become capable of transmitting them.

So, to state the question again, but this time more specifically: could AIDS be due to the combined result of under nutrition or malnutrition and environmental stress such as exposure to radiation and other contaminants? Further research is urgently needed to resolve the AIDS crisis.

How Blood Protects The Body

The blood combines respiratory, nutritive, excretory and immunoprotective functions. In performing all these functions, the blood carries various substances to the areas of the body where they are needed. Blood transports oxygen from the lungs to the tissues and carbon dioxide from the tissues to the lungs. It transports wastes to the kidneys. Nutritive substances, vitamins and hormones all circulate in the blood. And, as mentioned earlier, the blood produces antibodies, which, along with lymphocytes and other immunoprotective cells, fight infections throughout the body.

Our blood also protects us by maintaining the *pH balance,* which is the balance between acid and alkaline. This balance is crucial to the efficient transport of both nutrients and toxins. The pH scale measures hydrogen ion (H^+) concentration on a scale from 1 to 14. The lower the pH value, the more acidic the substance is; the value 7.0 is neutral. Ideally, the pH of our blood is slightly alkaline—about 7.4. The kidney and the respiratory system help to regulate the blood pH. Most significantly, however, the blood regulates its own pH.

Certain natural proteins buffer the acids. This fine-tunes the balance of life.

However, the alkalinity of our blood can be disturbed by our type of diet and our exposure to toxic substances. Excessive amounts of toxins or acid-forming foods such as meat and eggs can cause over acidity. This breaks down the proper balance between the electrical charge inside a cell and outside it, thus stopping all exchange of nutrients and waste.

The body may get rid of excess substances formed as a result of dietary imbalances and toxins in a similar way. It may set them aside either in the lymph nodes or in an organ, thus temporarily maintaining balance, but also setting the scene for later dysfunction and disease. In his book *Toxemia,* naturopath J. H. Tilden writes: "Every so-called disease is a crisis of toxemia, which means that toxins have accumulated in the blood above the toleration point and the crisis, the so-called disease—call it a cold, pneumonia, flu, headaches, or typhoid fever—is a vicarious elimination." Although the body may be healthy on the outside, dietary imbalances take their toll internally over a period of time.

Another naturopathic educator, Dr. Bernard Jensen, asserts: "First, the body must have a healthy bloodstream, for without this the body cannot have a healthy cell structure. Since cell life is dependent upon the blood, this must be kept clean and toxin-free." Like exposure to radiation, exposure to a strong and/or steady amount of chemicals leads to an increase in free radical reactions that create toxic substances which can upset the acid/alkaline balance of the blood. In addition, such exposure can damage the red blood cells so that they break. It also harms the bone marrow where red and white blood cells are formed, causing severe anemia and also reduced immunity due to a lack of white blood cells. A variety of nutrients are needed to build healthy blood. Chief among these are iron, folic acid, vitamin C, vitamin B_{12} and magnesium.

Carrying these thoughts a step further, I would like to share a proposal that I heard from a health counselor a few years ago. He suggested that we make a revolution of the bloodstream—to change our minds and attitudes by improving our blood. Throughout history, no group has consciously changed its diet and blood quality in order to make the individual body and mind more wholesome.

Out of this individual change will come global change. When we are not balanced we have no center and no center brings out fear. Changing blood quality will help us to move beyond fear and all its consequences in terms of world conditions.

How To Maintain Your Acid-Alkaline Balance

The radioprotective way of eating brings about a good balance of acid and alkaline in the body. Many studies have corroborated that a middle-range, slightly alkaline body chemistry enhances resistance to radiation. An article in the February 1980 issue of the *International Journal of Radiation Biology*, "The Influence of pH on the Survival after X-Irradiation of Cultured Malignant Cells," documented a definite influence by pH on the cell's response to radiation. They found that lowering the pH to a more middle-range value "resulted in higher survival after irradiation." This point is fundamental to evolving your overall body condition to be the most protective.

You may wonder how this could be possible. The mechanism is this: when the sodium and potassium ratio in the blood is 7 to 1, as many holistic health practitioners have found it should be ideally, bacteria cannot grow. This sets up an optimal system of preventative health and precludes many possibilities of ailments. It also provides the best transport medium for hormones to reach the glands and organs that require them.

In traditional Asian thought, foods are categorized as either acid or alkaline. This information is basic knowledge throughout the culture, because it is understood to be essential to good health. Very few Western doctors or nutritionists give this fact its due.

The blood of a person whose diet is balanced, including some acid-forming foods and some alkaline-forming foods, gets some exercise and enough rest, usually has a pH between 7.3 and 7.4 and is slightly alkaline. Stress causes the body fluids to become more acidic. When this happens, the body attempts to restore balance by neutralizing the acid. Calcium, an alkaline mineral, is reabsorbed from the bones into the bloodstream. This calcium loss can allow the inroads of strontium-90 (see page 66).

Americans are used to eating a lot of meat and poultry. In 1982, we ate nearly 250 pounds per person on the average, according to the United States Department of Agriculture. Acid-forming meat and poultry cause a desire in the body, one could say a craving, for balance. And so we develop a taste for sugar, sweet foods and alcohol, which are alkaline-forming. For example: sirloin steak and red wine; hamburger and apple pie; fried chicken and chocolate ice cream.

The radioprotective way of eating avoids foods that are either too acid or too alkaline and emphasizes *neutral* foods that are towards the center of the two extremes. Because grains are neutral, they help us to focus our energy. Vegetables and miso gently alkalize the body to offset the acidifying stress of modern life. The easiest way to balance acid and alkaline in your diet is to make sure that your daily intake of grains is about half of your total food intake, and vegetables about one quarter. Beans, sea vegetables, saltwater (ocean) fish and fruit are the best foods to round out the remaining one quarter. In this way the kidneys will have less work to balance the pH. You will feel better. You will feel more centered and satisfied. Radiation considerations aside, many potential ailments will be subverted.

A related problem worth discussing here is the often-heard complaint of stomach "acidity." In nine cases out of ten, the annoyance is probably really from a *lack* of hydrochloric acid, which is secreted by the stomach for the proper digestion of proteins. Many people are low in hydrochloric acid. Health researchers have found that at age sixty-five and over we have only about 15 percent of the amount we should have. Its lack is also stress-related. If proteins are not fully digested, toxic byproducts may remain and the condition may be construed as too acid a stomach. And then come the doses of the antacids that line the drugstore shelves—most containing toxic aluminum as the active substance. Beside its serious mental effects, aluminum figures in Alzheimer's disease and removes calcium from the body.

Usually a simple diet, lower in protein and higher in other nutrients, will restore low hydrochloric acid. Many of the foods in the modern diet—meat, eggs, dairy—are acid-forming, and without a balance of alkaline they weaken the constitution.

Traditional Eastern Views

Western medicine has analyzed the characteristics of acid and alkaline with respect to body physiology in great detail. Since radiation hazards know no geographical boundaries, however, we owe it to ourselves to extend our mental parameters and examine Eastern ideas as well. Eastern thought not only grants the importance of acid/alkaline balance to our health, but goes on to use the idea of complementary opposites, named yang and yin, to form the core of a vast, all-encompassing understanding of the many other dimensions of life. The fact is that environmental factors that Westerners don't often think of affect the body's balance.

Eastern philosophy says that yang and yin are constantly changing into each other. The *I Ching,* or *Book of Changes,* the cornerstone of Chinese wisdom, originated about 5,000 years ago and was required study for every scholar in China for 2,000 years. Based on an awareness of the constant flow of changes between opposites that maintains all nature, these teachings are as relevant to Western society now as they were to the ancient Chinese.

In *Cancer and the Philosophy of the Far East,* the modern philosopher George Ohsawa notes: "The source of all human tragedy lies in the demand for changelessness." A balance of yin and yang in our food choices will help us build a good blood quality, centering us. Then, on another level, Ohsawa feels, our true human judgment will awaken:

> Now scientific and technical changes have reached a dead end and a new and totally unknown avenue opens before us, leading to the development of man's judgment.
>
> All our behavior depends on judgment—madness or sanity, war or peace, happiness or unhappiness. But there are seven stages of judgment: mechanical, sensory, sentimental, intellectual, social, philosophical and supreme. The judgment responsible for scientific and technical civilization is limited to the first or second stage. We must now listen to higher sounds . . . .
>
> You have nothing to lose because you cannot lose what was never yours in the first place. Everything you call

"mine" will eventually be lost to you, for there is nothing permanent in this constantly changing world—nothing that is, except Change itself, the only constant . . . .

Finally, attain the seventh stage of judgment—Supreme, Infinite, Eternal Love where only endless happiness and Infinite freedom are seen, felt and known.

Eastern thought embraces the belief that all the characteristics of all things are determined by the proportion in which they combine yin and yang. This is especially apparent in our behavior and mental outlook. In *The Book of Macrobiotics,* Michio Kushi describes how the balance of yin and yang qualities in our food influences us:

When we include a larger amount of animal food than we really require, our mental activities tend to become more egocentric and aggressive towards the outer world. On the other hand, if we use vegetable food almost exclusively, with a large volume of fruits, we tend to exclusivity and a defensive attitude towards any strong stimulus coming from our surroundings. A large volume of yang, such as animal food, and a large volume of yin, such as salad and fruits, hot spices, and alcohol, produce fear and exclusivity, sometimes expressed similarly, but often with an opposite expression: the more yang category of food produces a more aggressive and offensive attitude, while more yin food produces a more defensive and self-excusing tendency. The former contributes more towards a materialistic view, the latter towards the forming of spiritualism. We should balance in the middle, avoiding excessive yang and yin qualities of food.

The way we eat is fundamentally important not only for our physical health but also for mental and spiritual balance.

5

The Foods to Avoid

There are two principles for protecting our internal environment from radiation and pollution. First, the *principle of optimal health,* which means keeping our cells saturated with the healthful nutrients they need to prevent the uptake of radioactive substances. Second, the principle of *avoiding detrimental foods*—foods that can limit our health (thus negating the first principle) and foster the absorption of radiation. (I call this second idea the principle of who-needs-them-anyway!)

These culprits are dairy products, wheat, meat and poultry, sugar, fats, fractionated (refined) foods and all foods processed using chemicals. Of course, dietary changeover should be gradual and also need not be fanatical. Radical sharp changes aren't good, so work with your current eating plan and reshape it to suit yourself. You can start by eliminating, reducing, avoiding or gradually cutting out—whatever is appropriate for you. What I will tell you here is for you to utilize to your advantage. For optimal health, what you eat regularly counts.

In the United States, we are not so strong in our health as we might think; about 25 million Americans have high blood pressure, 5 million have diabetes, one in three contracts cancer, and one in three men dies before the age of sixty from heart disease or stroke. All in all, the cost of health care in the United States is very high—$286 billion in 1981. Yet our current health crises are by no means unique. The other industrialized nations of the world face the same problems, which will undoubtedly have profound repercussions for generations to come.

Dietary Changes In This Century

There have been a number of significant dietary changes in this century. According to Michael Jacobson, author of the 1978 book *The Changing American Diet,* the average dietary fat consumption in the United States has increased by 31 percent and carbohydrate consumption has decreased by 43 percent since 1910. Today the average American gets 70 percent of his or her protein from animal sources, while at the turn of the century over half of our protein came from plants. Today over 60 percent of our calories come from fats and refined sugars, and only 20 percent from vegetables, whole grains and fruit.

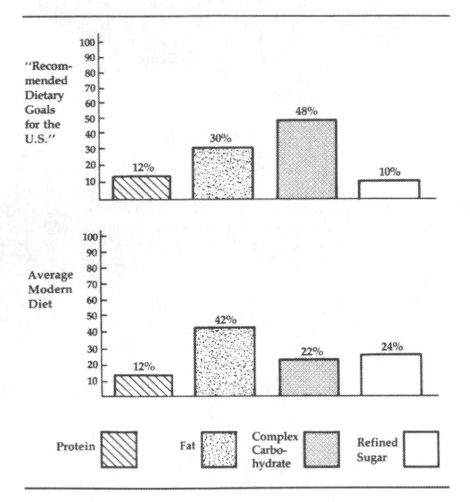

Average and Recommended Diets

Based on: *Dietary Goals for the United States*, 2nd ed. Prepared by the staff of the Select Committee on Nutrition and Human Needs, United States Senate. Washington: U.S. Government Printing Office, 1977.

An 80 percent decline in dietary fiber (see pages 126-129) corresponds to the decline of plant food intake. Unfortunately, this modern way of eating has opened the door for a vast increase in illnesses in this country. The "average modern diet" is illustrated in the graph on page 110. In 1977, a U.S. Senate Subcommittee concerned with the negative health effects of the typical diet recommended several changes. The recommendations are also shown in the graph.

Consider the fractionated foods we give our bodies. I believe these affect our thinking as we make decisions about our external resources. When viewed in this light, pollution, crime and the threat of war mirror our internal environment. In spite of all our technology, we haven't done so well.

Avoid Fractionated Foods

The road to optimal health and a balanced lifestyle is based on the consumption of *whole foods* from God's garden rather than the fast-food counter or the freezer chest. Whole foods are the opposite of fractionated foods. Only unprocessed, nutritionally complete foods without any chemical additives or preservatives qualify as whole foods. (Some healthful *natural foods* are partially refined using traditional methods.)

The intake of fractionated foods can cause blood cells to lose their electrical potentials (charges) and clump together. When this happens, they are less able to pass through the small capillaries where they usually deliver nutrients to the tissues. Also it becomes more difficult for the red blood cells to give up their oxygen and pick up carbon dioxide, a byproduct of metabolism that needs to be transported to the lungs, exhaled and replaced with fresh oxygen. Without the efficient transport of nutrients and oxygen in the blood, all metabolic and immune functions are lowered—including the activity of the liver, kidneys and bone marrow (where new blood cells are formed). As a result, the body's chance to protect itself from radiation is also limited.

Various modern methods of food processing add chemical preservatives and remove zinc, calcium and potassium, which are all imperative for radiation protection. These nutrients fill the body's requirements that would otherwise be filled by similarly structured radioactive elements (see chart on page 71). Food processing also removes the protective nutrients magnesium, vitamin B_6 and vitamin E.

For this reason, it is a good idea to skip artificial sweeteners, soft drinks, refined white flour products, canned fruit and vegetables, frozen products (including the not-so-time-saving "convenience" dinners) and all the other colorfully packaged products that line the supermarket aisles. In brief, avoid the four basic food groups—junk, packaged, fast and frozen. Their main contribution is to finance the manufacturers and inflate America's total health bill.

A point of interest: recent polls show that 96 percent of American children can recognize Ronald McDonald, the clown character associated with the giant hamburger chain. Among these children, he is second in public recognition only to Santa Claus.

Avoid Fatty Foods

Even the most conservative members of the medical establishment now recognize that the excessive consumption of fats in the diet can clog our arteries. Less widely publicized is the fact that fats in the diet also contribute to the formation of free radicals (see page 40)—resulting in damaged cells, weakened immunity and faster aging. The high fat intake of the average American (about 40 percent of total calories) is related to the incidence of obesity, strokes, heart attacks, diabetes and hypertension (high blood pressure) in our society. Reducing the likelihood of succumbing to these conditions is a benefit of the Atomic Age Diet; however, the focus here is on the protection of our internal environment from the external.

Researchers have found that fat cells attract and hold pollutants. In one experiment conducted by K. L. Davison, J. L. Sell and R. J. Rose, chickens were given a feed containing the pesticide called dieldren. Then their diets were severely restricted. As the body fat was broken down, as evidenced by weight loss, toxins that were held in the fat cells were released. The levels of dieldren in the blood rose rapidly, and some of the chickens died from dieldren poisoning. By holding pollutants and toxins in the body, fat hurts the immune system. When there are fewer fat cells to begin with, there are fewer pollutants in the body. Less fat, better immunity.

The terms "saturated," "unsaturated" and "polyunsaturated" relate to different molecular structures of fats. Butter is an example of a saturated fat, whereas margarine, vegetable oils, and oils from nuts and seeds are predominantly unsaturated. The role of saturated fats and cholesterol in coronary disease has received a lot of attention in recent years. Yet there are dangers from overconsumption of unsaturated and polyunsaturated fats as well.

These fats can affect cell division and endocrine (hormone) balance. They may also change the nature of the cell membrane so that it becomes easier for toxic substances to pass into the interior of the cell where they can damage the blueprints of the cell. All in all, unsaturated fats conspire to reduce the functioning level of the immune system. Thus, they may result in a variety of health problems, including cancer.

Should you use fats that are saturated, unsaturated or polyunsaturated? The body does need some fats to insulate nerve cells and ensure that metabolic processes run smoothly. The dilemma will be resolved if you generally avoid fatty foods. The necessary fats are easily obtainable from a diet containing whole grains, seeds and nuts, and unrefined vegetable oils.

An Update On Milk

Milk is a carrier of radioactive substances. I know it is hard to accept. Like so many American children, I was given a glass of milk at every meal, and I realize the confidence-inspiring aspects milk has. As we down a fresh cold glass of milk, we're assured that we are swallowing the All-American drink and meeting one-fourth our recommended daily allowance of calcium and protein. Yet milk is one of the foods that is best avoided.

On a daily basis, radioactive materials are emitted from nuclear plants. From the waste that has been used to cool the reactor, radioactive substances pass into nearby rivers, lakes or even the sea. They may contaminate water used for irrigating crops that are then fed to cows. As you may recall from our discussion of the food chain in Chapter 2, radioactive elements end up concentrated in many foods. Years of aboveground nuclear testing and American dietary habits combined led to an increase in strontium-90 uptake, especially during the 1960s. In 1967, a survey entitled "Strontium Metabolism in Man" verified: "Milk and dairy products are the main source of dietary Sr-90 [strontium-90] in the United States." In addition, like other fatty foods, milk and dairy products attract and hold contaminants. Apart from these concerns, however, there are other good reasons to avoid milk and milk products.

Estimates based on recent increases in milk production suggest that about two thousand years ago, a cow produced about 200 pounds of milk a year. Today a cow gives about 10,000 pounds of milk a year—thanks to modern science, which, with hormone injections, forced feeding and antibiotics has figured out how to turn this animal into a milk machine. The cow's life is shortened. It is more prone to ailments. And an interesting fact about the milk—it is thin and lacking in the nutrients it should contain.

Pasteurization is a process that raises the temperature of milk to sterilize it so that harmful bacteria are killed. Though they are neutralized, their remains are still in the milk. Pasteurization has other effects as well. Analyses of sterilized milk show that the B-complex vitamins and the mineral magnesium are destroyed to a certain extent. In addition, sterilization completely destroys one of the essential amino acids—making the other amino acids harder for the body to use. Butter, cheese, yogurt and ice cream are all made from milk, and share its nutritional defects and dangers.

Dairy products are also mucus-forming in the body. Excess mucus creates congestion in the bowels, lungs, bloodstream, lymph system and sinuses. Mucus also leads to the formation of acids in the tissues, diminishing our ability to ward off radiation. This is because increased acidity uses up needed B vitamins and causes calcium to leach out from our bones. We need calcium in our bones to prevent the uptake of strontium-90.

Perhaps the most telling fact is that milk from cows is made by nature for baby cows that grow fast, usually gaining about fifty pounds in the first month. The human baby hardly gains a pound a week at the same stage. Cow's milk has four times more calcium and three times more protein than human milk. Cow's milk fits the needs of the fast-growing calf.

The human baby, on the other hand, needs more carbohydrates and other nutrients for his brain and nervous system, which are still growing and developing up to the age of three. The composition of human milk is well-suited to the infant's needs. In all mammals except humans, the young avoid milk after weaning. And more than 25 percent of the world's population is allergic to cow's milk (and milk products) due to an inherited lack or deficiency of the enzyme lactose, which is necessary to digest it.

So—in spite of all the attractive ads on television promoting the use of milk as a healthful food, there is no reason to let ourselves be drawn into the use of a food that was made by nature for infants.

Avoid Sugar

On the average, Americans consume about 125 pounds of sugar per year. Some of this is consumed directly—three or four teaspoons in a cup of coffee, for instance—but the rest comes from packaged foods that contain hidden (or not-so-hidden) sugar. Canned pasta and canned vegetables generally contain sugar, as do catsup and salad dressing. Some types of gelatin are 90 percent sugar; popular brands of granola cereal have an average sugar content of 26 percent.

Table sugar, or sucrose, is refined from sugar cane or beets. Other forms of sugar include: lactose, the sugar from milk; maltose, the sugar from malt; and fructose, or fruit sugar, which is found in honey and fruit. All of these are converted into glucose in the body and used for energy.

Refined sugar contains no nutrients. Although it provides calories for energy, sugar is a prime example of what nutritionists call empty calories. And even the energy we get from sugar is short-lived, giving us a momentary feeling of well-being while causing havoc in the pancreas. The pancreas is the organ responsible for producing the right amount of insulin to keep the level of sugar in the blood steady. Refined sugars promote the overproduction of insulin, which reduces the sugar level too much. A slump in energy and mental function is inevitable.

Other ways sugar causes metabolic defects are almost too numerous to list—but here are a few. Sugar is acid-forming in the body. An acid condition can leach or draw the alkaline mineral calcium, which acts as a buffer, out of the bones. After a period of time, osteoporosis and kidney stones can result. In addition, sugar interferes with the body's use of B vitamins.

Abuse of sugar has consistently been associated with a variety of problems, including hyperactivity, shortening of attention span, learning disabilities and mental disturbances. Sugar has also been implicated in such diverse problems as arthritis, alcoholism, nervousness, insomnia, increase in blood fats, hypertension and the condition called *Candida albicans,* which may be manifested in lethargy, mental, confusion and chronic infections, among other symptoms.

The consumption of refined sugars exhausts the body; imbalances such as diabetes (high blood sugar) and hypoglycemia (low blood sugar) may result. In contrast, complex carbohydrates are used slowly and are ideal for the body's needs—offering a sort of timed-release sugar flow.

Sugar may be considered an absolute poison with respect to radiation. It makes the body more susceptible to its invasion. In his diary, published under the title *Nagasaki 1945,* Dr. Tatsuichiro Akizuki emphasized the importance of eating no sugar at all because it increases the severity of radiation sickness. Akizuki did not have any complicated explanations for why sugar reduced resistance to radiation damages, but in the aftermath of the bombing, his method made it possible for him to remain alive and to work ceaselessly on behalf of the many people who came to his clinic for help. Akizuki wrote:

> I had . . . no treatises about atomic disease. Yet I became convinced of the validity of my method of dietetics—mineral dietetics—which could simply be defined as follows: salt or a natrium [sodium] ion gave vitality to the haematogenic [blood-producing] cells, while sugar was their toxin.

Despite the fact that he says he did not have a very strong natural constitution, Akizuki survived the crisis and, in fact, is still alive today.

Today's health-conscious consumers are reaching for alternatives to simple refined sugar and are being tempted with such alternates as honey, barley malt syrup, maple syrup, fruit juice concentrates, amasake, molasses and fructose. The caveat here is to realize that each of these items is refined to a certain extent. I think of a "line of refinement"—a continuum ranging from items that are completely devoid of nutrients (like sugar and artificial sweeteners) to those that contain some beneficial elements (such as honey and maple syrup).

For optimal health, the thing to do is to remove sugar (along with some of the other old staples) from your shopping list. The gradual weaning from table sugar is not easy, because sugar is extremely addictive. Here, your discretion comes in—and the use of the less refined sugars when you just must have something sweet.

Avoid Refined Grains

Refined wheat, generally in the form of baked goods, is one of the greatest culprits as far as lowering our health level goes. Like other whole grains, when wheat is ground into flour and exposed to air, it loses much of its nutritive power. Because they are fractionated, the usual, easily available baked wheat products such as muffins, bagels, danish and croissants can cause mucus accumulation and congestion in the liver. Also, these items are baked with baking soda which contains aluminum, a toxic metal.

The refining of grains such as wheat and rice removes many of the radioprotective nutrients that are necessary for human health. Especially significant is the loss of vitamin B_6. Refining strips grains of 80 percent of their vitamin B_6. When grains are refined, their zinc content is also reduced. At the same time, refining leaves a higher proportion of the toxic metal cadmium, which can cause hypertension and kidney damage. For example, in whole wheat bread there is usually a zinc/cadmium ratio of about 35:1, but in refined white bread the ratio is 5:1. This can create problems because zinc, which helps to make a strong immune system and figures in other important body functions, also displaces cadmium. Chromium and other biologically significant trace minerals are also removed when grains are refined.

Refined grain products are "enriched" with vitamin B_1, vitamin B_2, niacin and iron. But vitamin B_{6+}, pantothenic acid (another B vitamin) and vitamin E are not restored. I suppose that this alone could account for a deficiency of these three nutrients, which are important for radiation protection. Vitamin B_6 is vital for the immune system (see Chapter 7 for an explanation of its unique value in promoting radiation resistance). Pantothenic acid keeps the adrenal glands from becoming depleted, a condition which would put a strain on the immune system. Refining removes 45 percent of the pantothenic acid. Vitamin E performs a variety of functions, of which perhaps the most outstanding is giving us protection from free radicals.

When the intake of refined, demineralized grains is coupled with a lot of sugar in the diet, it wreaks havoc in the body. The missing minerals zinc and chromium are needed by the pancreas to control blood sugar levels. Stress on the pancreas may result in either diabetes or hypoglycemia. The liver comes into play here also, as it is a storage site for sugars.

A diet high in refined foods stimulates the adrenal glands, which secrete hormones that prompt the liver to release sugar into the bloodstream. This stimulates the production of the hormone insulin in the pancreas. In this way, the body strives to regulate blood sugar. Depending on the amount of refined foods and sugars and how long we have been eating them, the liver, adrenal glands, and pancreas can deteriorate. Lower health levels, with manifestations such as allergies and early aging, are the result. Lower health, as we have seen, inhibits the body's power to resist radiation.

Avoid Meat

Is man a carnivore? To many people, eating meat is at odds with the idea of civilized human behavior. Apart from this philosophical issue, I would like to discuss the various things that determine the influence meat has on our health.

Assumptions about the diet of early humans have been reviewed and it seems that our ancestors were primarily herbivorous. Most probably, throughout the last million and a half years, dinner consisted of three times more plants than animal food—which is the *reverse* of the meal that appears on the modern dinner table! While mankind has consumed meat throughout history, the amounts we eat today are unprecedented. For example, the average American adult eats about 240 pounds of meat per year. I believe this has extreme consequences for the individual, the environment and the condition of our planet.

When we think about meat, we need to think about protein. The word *protein* comes from the Greek, and it means "to take the first place." The name is appropriate, because proteins are of prime importance for the processes of growth and repair. A protein molecule contains many amino acids, the first of which was discovered in the early 1800s. Since then, twenty-two amino acids have been isolated. Eight of these are considered essential. Meat, of whatever animal origin, is the single best source of essential amino acids. Thus, the modern meat industry has encouraged consumers to focus on meat in the diet to provide strength.

We need proteins, it is true, because they are the building blocks of the body, but enough is enough. Overkill is literally that. We should not ignore the strength of our ancestors as they ate combinations of grains and beans or vegetables which their instinct told them would provide the building blocks necessary for life.

The Food and Nutrition Board of the National Academy of Sciences recommended 120 grams of protein daily twenty years ago. By 1980, this recommendation had been lowered to 56 grams per day for adult males (and 44 grams per day for adult females). Some authorities advocate an intake of just 25 to 35 grams daily. In any case, the acknowledged needs have gone way down as our understanding of the body's needs has increased.

Recently, an excess of protein has been linked with many health problems, including over acidity, calcium-deficiency diseases such as osteoporosis, and buildups of toxic substances. In one research study, college students were given high-protein diets. The excretion of excessive calcium in the urine resulted, even when the students were given calcium supplements. As we have seen, it is important to maintain a high calcium level for radiation protection. In accordance with the principle of selective uptake (see Chapter 2), when our calcium reserves are low, our risk of absorbing radioactive stron-tium-90 increases.

Too much protein also puts a strain on the liver, which breaks down amino acids into usable elements. In turn, this stresses the kidneys, which have to filter the urea that is formed from the breakdown of amino acids. Urea contains ammonia, which is toxic. Excess protein—that is, above maintenance levels—increases the concentration of urea in the blood, creating an acidic condition.

In addition, meat is high in fat and calories. Like other fatty foods, it contributes to obesity. Moreover, it is an artery dogger. The arteries are pathways that must be kept clear so that the blood, carrying oxygen and vital nutrients, can circulate throughout the body. Arterial obstructions can cause high blood pressure, heart attacks and strokes; progressive hardening of the arteries due to fatty deposits may be halted through changes in diet.

Another problem is that meat is high on the food chain and can contain a high concentration of radioactive contaminants. An interesting point is that, unlike ours, the intestinal tracts of carnivorous animals are relatively short, so that they can quickly process meat without absorbing toxic substances. In contrast, it takes anywhere from several hours to over one day for meat, which is very low in fiber, to go through the human intestinal tract. During this time, it decays and putrefies, so we can't properly digest and assimilate the useful nutrients it originally contained. Instead, toxic byproducts are left in the body to clutter the bloodstream and lower our health. The slow intestinal transit time also can contribute to increased internal radiation absorption. So it is not surprising that there is a high correlation between meat intake and colon cancer.

Today's livestock suffer from internal pollution even more than we do, because they have no control over their environment. When we eat meat we must take account of the animal's toxic condition. Many factors contribute to this toxicity. The feed given to the livestock is exposed to chemical pesticides, fertilizers, and possible radioactive contamination. Antibiotics and synthetic hormones are added to induce abnormal growth and quicker fattening. Recently it has been estimated that between 12 and 15 million pounds of antibiotics—half of the entire amount used in the United States—are put into animal feed annually. This is in addition to any medications the animals may receive in the treatment of infections or disease!

When we eat meat we are also affected by toxic substances produced by the animal's metabolic processes—blood that was not purified by the liver, adrenalin that was released due to fear and uric acid that accumulated when the kidneys stopped processing waste.

Processing and packaging can affect the quality of meat. In March 1984, the United States Department of Agriculture admitted that meat it had been buying during a two-year period was contaminated with chemicals. According to an article that appeared in the May 1984 issue of *Vegetarian Times,* most of the 26 million pounds of meat involved had been distributed to school lunch programs. In September 1984, *The New York Times* reported that the owners of the meat packaging company had been found guilty of violating Federal laws. Testimony at the trial revealed that employees of the company had sneaked bad and rejected meat in with the rest after the Federal inspection. As many as 500 to 600 toxic chemicals may be in the American meat supply.

In addition to the hazards posed by all the various levels of contamination, the cooking of meat can also lead to health problems. Cancer-causing and mutation-causing substances are formed during broiling, roasting and barbecuing.

And there is one other problem when we eat meat. When we eat plant foods, they provide nutrients that build our blood and then our flesh. In the case of meat, however, plants eaten by the animal have become flesh, and then after being eaten by us are decomposed into substances absorbed by our blood. As we have seen, this decomposition is accompanied by the production of metabolic wastes and does not build high-quality blood. In a sense, the natural process of building the body is reversed.

Other Foods To Avoid

Besides the foods that are generally detrimental to our health, there are some others that can carry high levels of radioactive contamination under certain circumstances. It is important to be aware of this possibility.

After the Chernobyl accident, green leafy vegetables were thrown out in many areas. These were the ones that carried the immediate fallout. However, long-term fallout (such as cesium and strontium) gets absorbed into the soil and is then taken up by the deep roots of fruit trees and by some root vegetables (carrots, for example). Such long-term fallout can affect foods many years after a radioactive accident. In 1972, testing by the Florida Division of Health found that juice made from the fruit of Florida orange trees contained 190 picocuries of cesium per liter. For a point of comparison, consider that during the years of atmospheric weapons testing (prior to the Test Ban Treaty of 1963), cow's milk contained a much smaller amount of cesium—from 1 to 5 picocuries per liter. This does *not* mean that all carrots and other root vegetables, oranges and other fruits from deep-rooted trees contain cesium and need to be avoided. Many factors are involved, such as where you live and where the foods come from. Nevertheless, the potential for radioactive contamination of foods that in other circumstances would not be detrimental to health is a sobering possibility we must consider.

Another thing to consider is the vast difference in the radioactive content of freshwater fish and ocean fish. *Radio-nuclides in Foods,* a 1973 book prepared by the National Academy of Sciences Committee on Food Protection, Food and Nutrition Board, noted: "Levels of ^{137}Cs [cesium-137] ranging from 100 to nearly 5,000 pCi/kg [picocuries per kilogram] wet weight have been observed routinely in a variety of freshwater fish available commercially in the United States."

But the ocean contains minerals that fresh water either does not have, or has in much lower concentrations. Stable minerals in the ocean prevent the uptake of their radioactive counterparts by fish and other marine organisms. For instance, calcium in ocean water precludes uptake of the strontium-90 that would otherwise be absorbed. And, as noted in *Radionuclides in Foods:*

> The concentration of potassium in seawater is about 100 times higher than in fresh water . . . . This great difference in stable potassium levels in seawater undoubtedly explains the much higher concentration of radiocesium (a competitive sister element) in freshwater organisms.

Radiation is everywhere, but to minimize our intake of radiation in foods, it is a good idea to generally avoid eating freshwater fish, and to adjust our intake of fruits and root vegetables if they have been grown in an area exposed to inordinate amounts of long-lived radioactive fallout.

You Can Increase Your Energy And Look Better

Let me summarize the good "side effects" of leaving out meat and other detrimental foods and incorporating the ones discussed in the next chapter: a feeling of overall well-being; an absence of fatigue; a normalization of weight; an improved level of health; and last, but not least, monetary savings.

When you eat simply, the assured result is enhanced physical and mental energy—a clearer-thinking mind and a greater all-day vitality. A diet characterized by the absence of refined foods and an emphasis on whole grains, vegetables and beans is the way to a steady increase in stamina and sense of harmony.

For people who are committed to following a simple, natural way of eating, there is a gradual but definite and permanent leveling-out of the weight to what is ideal for the body frame. A 1986 survey shows over 30 percent of Americans are seriously (more than 20 pounds) overweight. Vegetarians, on the other hand, are found to be about 20 pounds under the national average weight. As a group, the vegetarians in our society are closer to their ideal weights. The large amount of complex carbohydrates (primarily from whole grains) in the vegetarian-type diet provides a lot of bulk with fewer calories but more nutrients. Complex carbohydrates satisfy hunger.

Most people who eat naturally will rarely be sick. Without a constant intake of mucus-forming foods and excess protein, the body's eliminative processes are allowed to "catch up." The accumulated wastes and mucus deposited around the body, especially in the sinuses, lungs and liver, will be gotten rid of. When you change your diet, though, you may experience some symptoms that are just good signs of the body's cleansing itself.

For example, a cold is the body's attempt to remove stored-up toxins by increasing elimination through the skin, lungs, nose, colon, and kidneys. If there is enough vitality and the eliminative process is not hindered by food or medicine (don't pay attention to what those TV ads say), the body is capable of restoring itself to health. When the body is congested, germs can multiply. I've always liked the succinctness of Louis Pasteur: "The microbe is nothing. The terrain is everything."

Chronic ailments result from repeated suppression of acute cleansing crises. The name given—be it arthritis, liver degeneration, or arteriosclerosis—is just a way of describing toxic substances that have accumulated in a particular location.

Now to the pocketbook. A week's marketing from the supermarket consisting of typical American fare is about twice the cost of foods from the health food store, local vegetable stand or farmer's market. So that's an improvement too.

The Diet for the Atomic Age is a balanced eating plan that does not clog the body with debris and allows the necessary continuing detoxification and elimination, while at the same time enabling the body to build new cells with high-level nutrients. Whole, natural, vital foods can create a vital body, and mind, and spirit.

Why not change your diet? What do you risk by cutting out meat, fat, sugar and refined and processed foods? (Do I hear you grumbling?) You may feel deprived at first, but as soon as you go through a period of transition the cravings fade out—and a new sense of well-being fades in. Your body is your creation. You can build high-energy cells which can resist radiation and environmental health hazards.

Revising "what most of us are used to eating" takes determination and planning ahead. But you have the satisfaction—and inspiration—of knowing that the quality of the food we eat does affect our health status.

6

A Menu for the Nuclear Age

Your health status affects your susceptibility to radiation. Exactly which foods should you eat and why? The foods that promote optimal health and create the most protective internal environment are whole grains, fresh vegetables, beans, sea vegetables, miso, tempeh and tofu, nuts and seeds. In this chapter, we will examine the special properties of these particular foods and the best ways to incorporate them into your daily diet. They constitute the key elements of the plan to use foods as a radiation shield.

How Foods Can Protect Us From Radiation

Many of the studies done to evaluate the cause-and-effect connection between food and health were gathered in one volume published in 1982 by the National Academy Press under the title *Diet, Nutrition, and Cancer.* The chapter "Inhibitors of Carcinogenesis" sums up the protective effects of foods: "Epidemiological studies have produced data suggesting that certain substances in foods may protect against the development of cancer." This study does not focus on low-level radiation. However, we know that low-level radiation causes a predisposing atmosphere for the development of cancer. Foods that create a more protective atmosphere are the ones that defend and fortify the internal environment. Protective foods promote optimal health in a number of ways.

We have already seen how antioxidants such as vitamin E get rid of free radicals before they can cause extensive cell damage. In addition, we have discussed the principle of selective uptake—how some nutrients, referred to as "blocking agents," displace their radioactive counterparts. Other nutrients bolster our immunity because they are essential for proper functioning of particular organs in the immune system (for example, zinc and vitamin B_6 for the thymus). The Diet for the Atomic Age provides all of these nutrients. In addition, the diet is rich in *cleansing* food factors, which chelate or bind with contaminants and facilitate their elimination from the body.

FIBER

Several naturally occurring components of food have the ability to chemically combine with toxic substances. Fiber is perhaps the most familiar of these cleansing agents. The others are phytates, sulfur-containing amino acids, sodium alginate and zybicolin. They are all important in the Diet for the Atomic Age. Fiber, for example, functions to protect us from radiation both directly and indirectly.

What Grandma called "roughage," food scientists now refer to as fiber. Fiber consists of the structural portions of plants, such as the bran of whole grains and the stems, seeds and skins of vegetables and fruits. Though it is neither digested nor absorbed in the human digestive tract, fiber is important to our diet because of its *effects*. It helps to regulate the way the body utilizes nutrients. In addition, it enables the body to maintain a continual state of detoxification, so that we become our own pollution-control device.

Since 1900, our dietary fiber content has declined about 80 percent. The intake of no other component of the diet has declined so drastically. There is no fiber in sugar or fat and very little in meat and most processed foods. A review of the literature was published in the *American Journal of Clinical Nutrition* in 1974. In this article, entitled "Antitoxic Effects of Plant Fiber," B. H. Ershoff expressed concern about these dietary changes in the Western World:

> Serious questions arise as to whether the ingestion of drugs, chemicals and food additives that may be without

deleterious effects when ingested by persons on high-fiber diets may not constitute a hazard to health for a substantial portion of the population in these countries.

One effect of fiber is to promote the growth of beneficial bacteria in the intestine. These bacteria synthesize the B vitamins, produce enzymes that improve digestion, and prevent harmful microorganisms form multiplying and producing toxins or carcinogens. This has a detoxifying effect and reduces stress on the immune system.

There are five main types of fiber. The insoluble fibers *cellulose* and *Hgnin,* found in beans, vegetables and the bran of whole grains, create bulk and speed up the passage of food through the digestive tract. In addition, cellulose acts like a sponge. It retains water and absorbs any toxic material that is dissolved in the water. Increased bulk and speed of waste elimination have a great deal of protective value, as they minimize our internal exposure to contaminants.

Pectins, gums and *gels* are soluble types of fiber from fruits, vegetables and legumes (peas and beans). By decreasing the absorption of fats in the stomach and small intestine, they offer the benefit of lowering fat and cholesterol levels. In addition, they slow the body's absorption of sugar. Thus, they "even out" the highs and lows in insulin production and blood sugar, and give us a more consistent energy level. It has been found that when some persons with diabetes switch to a whole grain, high-fiber diet under the supervision of their physician, they are able to go off insulin.

Many whole foods contain a mixture of fiber types. Others are noteworthy for particular types. For example, apples, grapes, cabbage and cauliflower are high in pectin; whole grains and beans are high in cellulose. Diets high in fiber have been shown to reduce the incidence of colon cancer, which claims more than 50,000 lives in the United States each year. Fiber also reduces the intestinal inflammation known as diver-ticular disease or diverticulitis, alleviates the symptoms of colitis or irritable bowel syndrome, and relieves constipation. In addition, research has suggested that fiber helps persons with heart trouble. All in all, fiber strengthens immunity and gives us a higher level of overall health.

Perhaps the most dramatic way in which fiber protects us from the effects of radiation, however, is through its *binding* ability. Lignins, gums and pectins chemically combine with poisonous substances. Once the molecules are bound together in this way, a new and much less toxic substance is created. Moreover, this new substance can be excreted by the body. And since fiber (particularly cellulose) also attracts and holds water, the poisons are further diluted and the waste material gains bulk that enables it to travel through the intestines quickly. The formation of chemical complexes that are readily eliminated by the body is also known as *chelation.*

How Are You Fixed for Fiber?

A breakfast of coffee and a donut, a lunch of soup and a ham sandwich, a dinner of roast beef, potato and gravy, and a dessert of chocolate cake can add up to as little as 3 grams of dietary fiber—a fraction of the amount we need. The human body has evolved over millions of years geared to handle high-fiber foods. We cannot adapt in a scant one hundred years (since food processing and refining first became widespread) to a diet in which there is virtually no fiber.

Dietary Fiber

Food	Grams of Fiber (per 100 grams)
Almonds	5.1
Corn	3.9
Apples	3.9
Lentils	3.7
Rolled oats	2.8
Broccoli	2.6
Barley	2.2
Kidney beans	2.2
Squash	2.2
Brussels sprouts	1.8
Kale	1.4
Brown rice	1.3

Note: Laboratories use many different methods to analyze the fiber content of foods. The values given in different reference books often vary considerably, reflecting these different measurement techniques. The above chart gives some widely accepted values.

Most authorities agree that we should have at least 10 grams of fiber a day. Dr. Dennis Burkitt, the researcher who brought the public's attention to the importance of dietary fiber, recommends 25 to 30 grams daily. And some nutritionists suggest that we may need up to 40 grams of fiber a day. The accompanying chart on dietary fiber shows how easy it really is to get the optimal amount of fiber. Simply include several portions of fiber-rich vegetables, beans and whole grains in what you eat each day.

The Protective Foods

The foods that protect and fortify our internal environment are natural, traditional foods. You probably are familiar with most of them—whole grains, fresh vegetables, beans, seeds and nuts—even if you don't eat them regularly. The paragraphs that follow outline the protective properties of these familiar foods, along with several that you may never have tried before—miso, tofu, tempeh and sea vegetables. You will see why these are all important to the diet.

Whole grains. Whole grains are high in complex carbohydrates, B vitamins, iron, zinc, calcium and trace minerals. They supply protein as well, but are low in sodium and fat. In addition, whole grains provide us with fiber and phytates, both of which are important against radiation. It is important to eat grains in their whole form, with only the inedible portions removed. The refining and processing methods most widely used today (for example, in converting whole wheat into white flour) seriously undermine the nutritional and radioprotective qualities of grains. Even when refined flour, pasta and other grain products have vitamins and minerals added, they are still fractionated foods. Fortunately, many different whole grains are available today. They include brown rice, millet, barley, corn, buckwheat, wheat, oats and rye. For optimal health, at least half of every meal should be grains.

Vegetables. Fresh vegetables are also a good source of fiber. They supply calcium, iron, vitamin A, vitamin C and the B group as well. The sulfur-containing amino acids cysteine and methionine are found in some vegetables. Approximately one-quarter of each meal should consist of vegetables. Leafy greens, yellow vegetables and vegetables in the cabbage family are all important in the Atomic Age Diet.

Beans. Beans are a concentrated source of vitamins, minerals and protein. The protein in beans is especially important because it *complements* the protein in whole grains; that is, it supplies the amino acids that are often lacking in grains. In addition, beans provide dietary fiber. Lentils, aduki beans and chick peas are excellent choices for daily consumption. Split peas, pinto beans and other varieties are suitable for occasional use. About 5 percent of the foods you eat each day should include beans.

Miso. One of the most efficacious protective foods is made from fermented soybeans. Miso is noteworthy as a vegetarian source of vitamin B_{12}. A soup containing miso, wakame sea vegetable and a fresh vegetable (with other ingredients as desired) is good once or twice daily.

Tofu and tempeh. Tofu is a traditional Oriental food, similar in texture to a light cheese. This soybean product supplies complete protein and many nutrients. Tempeh is also a traditional food made from soybeans, but is fermented. Like miso, it is a vegetarian source of vitamin B_{12}. Both tofu and tempeh are low in fats and calories and are versatile alternatives to meat.

Sea vegetables. Sea vegetables are a good source of minerals from the sea, including iodine. Sea vegetables also contain sodium alginate, a valuable chelating agent. Sodium alginate chemically binds with radioactive substances and toxic heavy metals, thereby converting them into salts that the body can eliminate. Sea vegetables should comprise about 5 percent of the daily diet. The sea vegetable kombu can be used often in making soup stock.

Note: Caution on soy foods and sea vegetables. Consider reducing grains and increasing vegetables.

Seeds and nuts. Sunflower seeds are high in pectin, a type of fiber that binds toxins. Sesame seeds have a high calcium content and their use with rice provides a complementary protein.

Almonds are a source of minerals and have traditionally been regarded as especially health-building. The soy products, seeds and nuts should make up about 5 percent of the diet.

While the protective foods listed above are the key elements of the Menu for the Nuclear Age, you need not limit yourself to them. Several times a week a small amount of seafood may be eaten, preferably a small, white-meat type of fish. In addition, a modest amount of locally grown fruit is recommended. Of course, it is important to make sure that you obtain fish and fruit that have not been exposed to inordinate levels of environmental contamination.

Herbs and spices can enhance any meal. The most protective ones are ginger, horseradish, scallions and garlic (see Chapter 7 for information about garlic in supplement form). Some herbs, such as red clover, also have cleansing properties.

Keep the principle of selective uptake in mind when you decide what to eat each day. If you have enough of a needed element, your body won't grab onto a radioactive lookalike in its avid pursuit of maintenance.

It is easy to feel discouraged and overwhelmed by the prospects of future nuclear accidents. But that sense of futility, which says "Why-bother-there-is-nothing-to-do," need not oppress us. We can dismiss it. By watching our diets, we can do all we can for ourselves and those close to us. Those of us who maintain the highest state of health will be among the fittest—the survivors.

Whole Grains

In his book *The Staffs of Life,* E. J. Kahn points out: "Most people on earth do not eat much peanut butter and jelly, or much beef or pork or fowl or fish, though they probably wish they could. They eat rice and wheat and corn and sorghum and millet and cassava and potatoes. These, to them, mean food—the rooted staples without which, for more than half the world's population, life would be unsupportable. Contemporary farmers can readily produce enough grains and tubers to provide every living human being with three thousand calories a day of nourishment—well above the accepted minimum for an adequate diet."

Nevertheless, according to the United Nations' Food and Agricultural Organization (FAO), about 500 million people go hungry in the world today. The modern world has largely discarded whole grains as the staff of life, in favor of fast foods with an antacid chaser. All ancient peoples knew that grains are the staff of life. The early civilizations of Egypt, Greece and Rome were based on millet, barley and wheat. Rice was the staple grain in India, China and Japan. The Maya, Inca and other Central and South American cultures lived on corn. Throughout eons, human beings have flourished on a diet of natural foods, mainly grains, vegetables, nuts, tubers and fruits, with some meat or fish on the side.

An archaeologist studying what was eaten in ancient China found millet at some archaeological sites dating to about 5,000 BC. I think of that when acquaintances see me eating grains and ask "Why are you always eating some new-fangled health food?"

But interest in grains is coming back. The American Cancer Society is currently proposing a diet of whole grains and vegetables. Over the last few years, the Nutrition Subcommittee of the U.S. Senate, the U.S. Department of Health and many leading agencies have focused attention on the value of whole grains and vegetables.

Most of the food energy we get from whole grains and vegetables comes in the form of *complex carbohydrates,* in contrast to the proteins, fats and refined sugars that predominate in other foods. The two bestselling health books of 1985 reaffirmed the value of complex carbohydrates. In *Eat to Win,* Dr. Robert Haas states: "Complex carbohydrates are the best foods for peak performance because they are the only truly clean burning, readily available source of blood sugar. Complex carbohydrates such as brown rice and pasta are worth their weight in gold to professional athletes . . . they are no less valuable to weekend athletes who want to excel at their favorite activities." *Dr. Berger's Immune Power Diet* states: "The complex carbohydrate foods such as potatoes and whole grains are a key . . . . Like sugars these foods are high in energy, but unlike sugar, they break down slowly in our bodies. In effect, they work as time release energy sources."

I used to think grain was something you poured chicken gravy over, or the stuff that sat next to the duck a l'orange. But, enjoying a variety of grains over the last few years, I've realized that, instead of an accompaniment, they should be the main course.

Grains are actually seeds, and so they contain protein and concentrated nutrients needed to start a new plant. They are high in B vitamins, fiber and minerals, including chromium, which helps prevent diabetes and hardening of the arteries. Whole grains supply us with complete protein (that is, the eight essential amino acids) when eaten in combination with beans, tofu or fish. And they are low in fat.

Whole Grains Protect in *Five* Ways

Whole grains help to protect us from the deleterious health effects of radiation exposure in five ways:

1. Grains are low on the food chain. Although they may have been exposed to pollution and radiation, they do not have the concentration of contaminants that is found in meat and large fish, which are at the top of the food chain.

2. Important with respect to radiation protection is the high fiber content and phytates in grains. The binding ability of these substances helps the body to remove poisons.

3. The bulking factor of grains lessens the intestinal transit time and so hastens the elimination of all toxins.

4. Being neither very acid nor very alkaline, grains help us to maintain the middle-range pH that has been found to increase our resistance to radiation.

5. Whole grains provide vitamin B_6, which is indispensable for the thymus. In addition, their calcium content guards against uptake of radioactive strontium, and their vitamin E and selenium prevent cellular damage caused by free radicals.

Grains and the Food Chain

Diet for a Small Planet, a paradigm-altering book, appeared in 1971. Frances Moore Lappe presented a case for grains in terms of the food chain:

> My purpose is to show you a way to minimize the amount of ecologically concentrated pesticide and heavy metal you ingest: by eating low on the food chain, you are simply reducing the quantity of most if not all pesticide residues in your diet.

> To the two poisons she names we can add low-level radiation.

Whole Grains Offer an Easy Way to Lose Weight

Because high-fiber foods are bulky, we are less inclined to overeat. Whole grains provide us with nourishment and satisfaction that make it easier to lose weight. This is especially true when combined with the elimination of *all* baked flour products. Refined grains are low in fiber and do not provide a satisfying balance of nutrients. Perhaps because it is so easy to eat too much of these fractionated foods, many people believe that carbohydrates are fattening. But as the innovative nutritionist and health educator Nathan Pritikin said: "Much maligned carbohydrates turn out to be not only the healthiest kinds of foods we eat, but also the kinds that keep people slim." The key to weight control is to base your diet on the carbohydrates that come the way nature packages them—in whole grains.

Neutral pH

When digested and metabolized in the human body, grains have a beneficial effect on body chemistry. They are neither too acid nor too alkaline. Rather, they have a neutral pH (that is, close to 7.0, in the middle of the pH scale). Millet is the only slightly alkaline grain. The others are just slightly acidic. All are in the middle range.

Neutral pH has been shown to limit radiation effects. A 1980 article in the *International Journal of Radiation Biology,* "The Influence of pH on the Survival after X-Irradiation of Cultured Malignant Cells," described this decrease in radiation sensitivity with a middle-range pH.

Going with the Grain

Millet is the only complete protein grain. It is comparable to meat in that it contains all the essential amino acids. Light and

How Whole Grains Protect

Nutrient	Source	Function
B-complex vitamins	all whole grains	help nervous system and immune system; essential for utilization of food energy
vitamin B_6	all whole grains	essential for thymus; helps build blood
vitamin E	all whole grains	antioxidant, gets rid of free radicals; aids heart and circulation
calcium	all whole grains	strengthens bones; blocks uptake of strontium-90
chromium	all whole grains	helps stabilize blood sugar
iron	all whole grains	builds red blood cells; blocks uptake of plutonium
magnesium	all whole grains	helps maintain pH balance
selenium	all whole grains	antioxidant, gets rid of free radicals
zinc	all whole grains	essential for thymus; blocks uptake of zinc-65
fiber	all whole grains	binds with radioactive and toxic substances
phytates	all whole grains	binds with radioactive and toxic substances

pleasant-tasting, millet makes a wonderful breakfast cereal with raisins and almond milk. It can also be used in many recipes in place of rice.

Brown rice is best served with beans to complement the missing amino acids. Brown rice is chewy and can be very tasty; serve it daily. Experiment with short grain brown rice, which is more sticky, and long grain, which turns out a separated fluffy grain. Try a mixture of two-thirds rice and one-third barley. Rice is especially high in the B-complex vitamins; thus, it has a calming effect on the nervous system.

Buckwheat, or kasha, is another power-packed grain, special in that it contains rutin, a substance that is helpful for the capillaries and aids circulation. Manganese, magnesium and vitamin E are also found in a good amount. Buckwheat is delicious many ways: as a cereal, cooked in various dishes (as you would use rice), or made into buckwheat noodles. It is a good food in the colder months, as it is very warming. In addition, it is supposed to be beneficial for the kidneys.

Another good grain for the winter is oats. In addition to whole oats, fresh steel-cut oats are very good. Whole oats need to be soaked overnight before cooking or cooked an hour, but are worth the effort, as they make a splendid winter breakfast. Oats contain more fat than other grains. Although there is little hard evidence on this point, they are supposed to be supportive for people with an under-functioning thyroid gland. In any case, oats are rich in iron and calcium.

Barley is one of the oldest cultivated grains. It is easy to digest and mixes well with other foods. Currently, most of our barley crop is used for animal feed; about one quarter is used to make beer and malt whiskey. But barley is also famous as barley soup and can make a delicious casserole with onions and other vegetables.

Rye is similar to barley in that it is more often drunk by Americans than it is eaten! This grain yields a dense and heavy flour, so it is combined with other grains to make it more palatable. Popular in Europe, rye is tasty and quite chewy. It has a reputation for building stamina and energy.

Corn was eaten by the Maya and early Inca, and was a staple of North and South American Indians. Today it is often utilized as cornmeal or the more finely milled corn flour to make delicious bread and muffins. Corn has traditionally been regarded as a blood builder.

Vegetables

How many times, growing up, did I hear "Eat your vegetables"? I don't have to wonder now why my parents had to insist—those vegetables were cooked in too much water for too long and the result was flavorless. In contrast to this, I'll never forget the first time I cooked beets and carrots freshly picked out of my vegetable garden—what flavor and sweetness! Now I don't have a garden, but I make an effort to find vegetables as fresh as possible and I steam them for just a short time.

But in urging me to eat my vegetables, my parents were right about one thing. Vegetables do have a great value in helping us to maintain a strong body chemistry. They protect us from cancer and the effects of radiation. Sulfur-containing amino acids in vegetables bind with toxic substances so that the body can excrete them. Vegetables also build the blood, the thymus and immunity. As they contain so many nutrients, they fortify the body in accordance with the principle of selective uptake. They help us to defend against bacteria and protect our cells. They are important for fiber as well.

Eat Green Vegetables for Red Blood

Green plants contain chlorophyll, a substance that is rich in the mineral magnesium. In a sense, chlorophyll can be called the bloodstream of the plant. Green plants transform light energy into chemical energy for human consumption. With the use of the energy of sunlight, plants convert water from the soil, along with carbon dioxide from the air, into starch and protein. Thus, when we eat plants we are consuming sunlight in another form. In addition to containing chlorophyll, green vegetables are rich in blood-building iron and vitamin C.

Vitamin C, which is found in green vegetables, does a great variety of jobs—it builds blood, it counteracts toxic substances, it aids the adrenal glands in dealing with distress, it supports the immune system, and it has been found to protect cells from radiation effects. Overall, vitamin C plays a major role in detoxifying the body. At least one study has found that cells do not become cancerous after exposure to x-rays when vitamin C is present. It is important to eat vegetables when they are fresh, because their vitamin C content diminishes rapidly as time passes.

High-quality blood has radioprotective powers. It is better to build your blood with iron-rich leafy greens, instead of with iron supplements, which are only a partial aid to the making of rich red blood. Folic acid, vitamin C, vitamins B_6 and B_{12} and magnesium (which is contained in chlorophyll) are also necessary.

Our bodies are constantly renewing themselves. After ten to fourteen days or so, most of the white blood cells circulating in the bloodstream are old. In an ongoing process, such cells are removed from the bloodstream and replaced with new cells. Virtually all of our white blood cells are replaced every one to three months. Red blood cells are replaced every four months. Over a period of time the quality of the bloodstream changes when you eat leafy greens. You really can transform yourself.

Calcium from Leafy Greens

Blood building is not the only benefit that leafy greens offer. In addition to providing chlorophyll, iron and vitamin C, leafy greens are rich in: B-complex vitamins, particularly folic acid; vitamins A and E; and the minerals potassium, magnesium and calcium.

In fact, I've noticed that people who have grown up with greens as a calcium source rather than milk generally have very white teeth—a sign of excellent nutrition. Leafy green vegetables such as turnip greens, bok choy, kale, collards and watercress are particularly good ways to take in calcium which the body can use well. Other leafy greens that contain calcium include spinach, swiss chard and beet greens, but the calcium from these vegetables may be more difficult to assimilate, due to the presence of oxalic acid.

Yellow and Green Vegetables

A number of epidemiologic studies have shown that the risk of many types of cancer is directly related to the amount of green and yellow vegetables eaten. In general, the more of these vegetables you eat, the lower your chances of developing cancer! It might stretch your credibility to hear that carrots can be a potent protector, so let me quote for you from the journal *Science* regarding the *carotene* in carrots:

Carotene is another anti-oxidant in the diet that could be important in protecting body fat and membranes against oxidation. Carotinoids are free radical traps and remarkably efficient quenchers of singlet oxygen, which is mutagenic. Carotene is present in carrots and all food that contains chlorophyll.

Carotene is what is known as a vitamin A precursor. In other words, the carotene in the vegetable is converted to vitamin A in the body. Vitamin A maintains the skin and mucous membranes, which are a stopping point for bacteria and pollutants. Vitamin A fortifies the thymus and immunity. And vitamin A is an antioxidant that is very effective against free radicals. Along with carrots, squash, corn and parsnips are rich in carotene. So are kale, spinach, green beans and many green vegetables in the cabbage family. It is important to be aware of these alternative sources of carotene, because if soil is contaminated with cesium-137, carrots (and other root vegetables) are likely to absorb harmful radiation. When this is the case, it is best to rely on other green and yellow vegetables for carotene.

The *American Journal of Clinical Nutrition* published the results of a study monitoring 1,200 residents of Massachusetts, calculating their intake of vegetables, and found that those who ate the most vegetables had the least cancer risk. The researchers hypothesized that the carotene content was the protective factor, but recognized that it might be something else.

Vitamin B$_6$, which all too often is refined out of grains, is found in green leafy vegetables, cabbage and carrots. As discussed in Chapter 4, it is crucial for the thymus and immune function.

The Cabbage Family

Vegetables that contain sulfur have been found to have strong protective power against radiation. They include: broccoli, brussels sprouts, cabbage, cauliflower, chard, kale, mustard greens, onions, parsley and watercress. The sulfur is contained in amino acids that are found in these particular vegetables (other sources are fish, eggs and meat).

Most amino acids contain only carbon, hydrogen, oxygen and nitrogen, but *cysteine* and *methionine* also contain sulfur.

These amino acids function as antioxidants, free radical deac-tivators and poison neutralizers. They have a high affinity for radioactive substances and for toxic heavy metals. They bind or complex with them; in this form, the poisons can be excreted in the urine.

In his popular handbook, *The Herb Book,* naturalist John Lust explains this process in simple terms: "Sulfur . . . helps to counteract toxic substances in the body by combining with them to form harmless compounds." And, as the pioneering health educator Adelle Davis noted in her 1965 book *Let's Get Well,* sulfur amino acids help maintain the liver, which in turn does detoxifying work.

In 1984, the American Cancer Society recommended that Americans eat more of certain specific foods—including cabbage, brussels sprouts and broccoli. Botanically speaking, many sulfur-containing vegetables are members of the *Cruciferae,* or cabbage family (see list below). They are sometimes referred to as *crucifers.*

The Cabbage Family

Bok choy	Kale
Broccoli	Kohlrabi
Brussels sprouts	Mustard greens
Cabbage	Radishes
Cauliflower	Red cabbage
Chinese cabbage	Rutabagas
Collards	Turnips
Horseradish	Watercress

In 1950, the researchers Lourou and Lartigue published their observations on the relationship between diet and radiation in the journal *Experientia.* They fed one group of guinea pigs cabbage and another group beets; then they exposed both groups to a dose of radiation. The cabbage group had lower mortality. Later, a similar experiment was conducted by Spector and Calloway. One group of guinea pigs was fed just grains and the other grains and cabbage. These experimenters found 52 percent less mortality in the cabbage group. Calloway and others repeated the experiment to test the rest of the cabbage family, including brussels sprouts, and obtained similar results.

Several articles I found in medical journals explain how sulfur compounds deal with radiation on the cellular level. For example, the author of "Mechanisms of Radioprotection—A Review," discusses the idea that sulfur forms a short-lived chemical complex with the *mitochondria.* These microscopic structures, scattered throughout the cell, are the sites of cellular respiration—they break down nutrients and release energy in the process. The article states that "binding of the sulfur compounds primarily to mitochondrial membranes" makes the cellular system become "temporarily radio-resistant." Thus, it seems that one benefit of consuming cabbage-family vegetables is that more sulfur will be available to the mitochondria.

Another study, "Mechanism of Action of Aminothiol Protectors," finds that sulfur also benefits the cell nucleus: "There is ample evidence to show that DNA is the site of primary radiation damage in cells. Sulfur helps to repair the DNA." The August 1970 article "Free Radicals in Biological Systems," published in *Scientific American,* attributes sulfur's role in cellular repair to its effect upon harmful free radicals: "Since sulfur groups react readily with radicals, it is not surprising that sulfur compounds act as drugs that protect against radiation. There are a number of mechanisms by which a molecule might protect a cell from radiation. For example, compounds containing S-H groups (thiols) [sulfur bound to hydrogen] can protect important biological molecules through a repair process." The result is a "less lethal" type of radical.

Besides being radioprotective, the cabbage family can also help us to decrease our cancer risk. An article by Lee Wattenberg *et al,* published in 1976 and entitled "Dietary Constituents Altering the Responses to Chemical Carcinogens," was one of the first to appear. Finding that there are compounds that inhibit the cancer-causing effects of chemicals, the authors note: "An increasing number of these . . . are being found in natural products. Cruciferous vegetables, including brussels sprouts, cabbage, and cauliflower contain such compounds." They also mention turnips and broccoli. In January 1985, Anderson, Pantuck *et al* reported that research involving human volunteers found "significant ways in which components of the human diet can alter patterns of metabolism and the overall impact of chemicals on humans." Brussels sprouts and cabbage, for

How Vegetables Protect

Nutrient	Source	Function
vitamin A	carrots, corn, green beans, kale, leafy greens, parsnips, squash, zucchini	maintains skin and mucous membranes; helps defend against infection; antioxidant, gets rid of free radicals
B-complex vitamins	all vegetables	help nervous system and immune system; essential for utilization of food energy
folic acid	green vegetables	helps build red blood cells
vitamin B_6	leafy greens, cabbage, carrots	essential for thymus; helps build blood
vitamin C	green vegetables	helps build blood; counteracts pollution; aids adrenal glands; supports immune system; antioxidant, gets rid of free radicals; overall detoxifier
vitamin E	leafy greens, vegetable oils (unrefined)	antioxidant, gets rid of free radicals; aids heart and circulation
calcium	leafy greens	strengthens bones; blocks uptake of strontium-90
iron	leafy greens	builds red blood cells; blocks uptake of plutonium
magnesium	green vegetables	helps maintain pH balance
potassium	all vegetables	helps regulate pH of body fluids; blocks uptake of cesium-137

sulfur	broccoli, brussels sprouts, cabbage, parsley, watercress	resists radiation on cellular level; helps repair DNA; counteracts toxins; blocks uptake of sulfur-35
zinc	spinach, green peas	essential for thymus; blocks uptake of zinc-65
fiber	all vegetables	binds with radioactive and toxic substances

example, have been found to enhance the body's use of certain pain-killing medications. Believe it or not.

Beans

You may remember shelling green peas out of a pod. Such oblong pods containing seeds are characteristic of the *legume* family. The legumes include peas and beans. For the sake of simplicity, however, I will refer to these foods collectively as beans. In addition to green peas, other fresh beans include green beans, yellow beans and lima beans. Some beans are primarily available in dried form. These include aduki, pinto, kidney, navy, great northern, split peas and soybeans. Beans are an important source of protein and calcium. They also provide B vitamins and vitamin A. Moreover, they contain a variety of cancer-inhibiting and radioprotective factors.

The tradition of eating beans with grains is found in most parts of the world. In the Orient, beans or bean products accompanied the staple rice; in the Middle East, wheat bread and hummus (a paste made from chick peas and spices) satisfied generations; and in our part of the world, North and South American Indians combined corn and beans. Somehow, ancient civilizations knew instinctively what we have just recently discovered: grains and beans offer complementary proteins. The essential amino acid that is missing in one food is found in the other. Beans are rich in lysine, which is lacking in grains. These complementary proteins do not necessarily need to be eaten at the same meal. As long as your diet is well-balanced, and adequate on a day-to-day basis, the amino acids will be balanced out of the body's protein pool.

Radioprotective Factors in Beans

Factors found in beans have been found to eliminate radioactivity. *Phytates,* including phytic acid, are phosphorus compounds that are found in most plant foods but are especially abundant in beans, peas and whole grains. Phytates have the property of combining with toxic and radioactive elements and forming compounds that are eliminated via the intestines. A 1980 report by the National Council on Radiation Protection, "Management of Persons Accidentally Contaminated with Radionuclides," noted:

> Phytates, as found in grains, particularly oats and soybeans combine with radioactive substances and excrete them.

The fact is, phytates are a binding or chelating agent. They function the same way fiber does. Since beans contain fiber as well as phytates, their radioprotective value is enhanced.

But beans also have other protective factors that function in a different way. These are *protease inhibitors,* which are present in beans and seeds. Although their biochemical role is not completely understood, the main function of protease inhibitors is to prevent metabolism of proteins that might be injurious. In the 1970s, investigators began to realize that protease inhibitors impede the formation of cancers and also lessen radiation damage to biological systems.

At the New York University Medical Center, a biochemist named Walter Troll did extensive work with protease inhibitors and found that soybeans, lima beans and other beans and seeds protected against the growth of cancerous tumors in laboratory mice. He reported his findings to the American Cancer Society in March 1983, stating that when lab animals were fed a diet high in protease inhibitors they had a much higher resistance to carcinogenic substances.

Ann R. Kennedy and John B. Little, the authors of a 1981 study entitled "Effects of Protease Inhibitors on Radiation Transformation *in Vitro"* investigated the ways that three different protease inhibitors affected the extent of the damage ("transformation") that x-ray exposure caused in mouse cells. They found varying degrees of protection—from antipain, which suppressed radiation's alterations completely, to leupeptin, which was less effective, to soybean trypsin inhibitor, which only suppressed certain effects.

The Bowman-Birk, which is isolated from soybeans, is another typical protease inhibitor. It was chosen by researchers to observe the mechanism by which protease inhibitors offer protection. A study published in *Cancer Research* in May 1983, "Bowman-Birk Soybean Protease Inhibitor as an Anticarcinogen" outlined how protease inhibitors protect us: first, by causing a decreased absorption of proteins, and second, by counteracting the formation of free radicals. Authors J. Yavelow *et al* concluded:

> Protease inhibitors . . . can escape the usual digestive route of proteins and appear fully active in the duodenum [small intestine] where they inhibit the absorption of other proteins by complexing with proteases . . . . Thus, decreased absorption of proteins may be in part responsible for the anticarcinogenic action of protease inhibitors.

Noting that "ionizing radiation is known to form oxygen radicals directly from water which contributes to tissue injury," and that "the radioprotective effect of protease inhibitors on ionizing radiation has been observed previously," the authors also stated:

> The finding that BB [Bowman-Birk] is able to specifically block the effect of ionizing radiation can be interpreted as an effect on oxygen radicals by this protease inhibitor.

There is an important point to be made here, however. In order to derive a benefit from the radioprotective effects of beans, it is important to maintain a state of general good health and nutrition. This is because while phytates complex with toxic substances, they can also combine with needed minerals such as calcium and zinc, making them less available to the body. Similarly, while protease inhibitors can decrease the absorption of injurious proteins, they can decrease the overall availability of protein. Fortunately, however, research has indicated that the human body has an adaptive response. When we become accustomed to whole foods, our bodies become more efficient at extracting needed nutrients. It is for this reason that sudden, sweeping dietary changes are not recommended. The slow introduction of foods you are not used to is suggested to gradually build your level of health.

Beans in the Daily Diet

Even once you understand that beans are extremely beneficial, you may still say they are indigestible! The substances in beans known for producing intestinal gas are called oligosaccharides.

How Beans and Soy foods Protect

Nutrient	Source	Function
vitamin A	most beans	maintains skin and mucous membranes; helps defend against infection; anti-oxidant, gets rid of free radicals
B-complex vitamins	most beans, miso, tempeh, tofu	help nervous system and immune system; essential for utilization of food energy
vitamin B_{12}	tempeh, miso	helps build red blood cells; essential for nervous system; blocks uptake of cobalt-60
calcium	most beans, miso, tofu	strengthens bones; blocks uptake of strontium-90
iron	all beans, miso, tofu	builds red blood cells; blocks uptake of plutonium
potassium	all beans	helps regulate pH balance of body fluids; blocks uptake of cesium-137
zinc	soybeans	essential for thymus; blocks uptake of zinc-65
fiber	all beans	binds with radioactive and toxic substances
zybicolin	miso	binds with radioactive and toxic substances

They are broken down by a process of fermentation that results in gas. By soaking beans before cooking them and changing the water part of the way through the cooking time, it is easy to leach out this gas-producing material, leaving a delicious and digestible and dependable food.

There are many kinds of beans. Aduki beans, lentils and chick peas are among the most nutritious. Also try others for occasional use—for instance, split peas, kidney beans, lima beans, pinto beans and soybeans. A small amount is all that is needed daily—about two tablespoons per serving.

To improve the digestibility of dried beans, it is necessary to soak them in a bowl of water for several hours (or overnight). If the temperature is warm, put the bowl in the refrigerator. Discard the water after soaking the beans; when you are ready, cook them in fresh water for about thirty minutes. Then drain the beans and discard the water. Finish cooking with fresh water. Slow cooking over a low flame also improves the texture, flavor and digestibility of beans.

The one type of bean that does not need soaking is lentils, which cook quickly. In forty minutes, you can make a great lentil soup that can serve as a hearty lunch or dinner. When you cook beans of any kind, it's a good idea to make extra and keep the leftovers in the refrigerator. You can add them to soup later or serve them for lunch the next day.

Miso Is A Power Food

Miso is a paste made from soybeans and sea salt, often mixed with a grain such as rice or barley. A naturally occurring bacteria causes it to ferment and produce enzymes. Miso is aged an average of eighteen months. Making it is considered an art because of the delicate influence of the proportions of the ingredients, temperature and humidity during the fermentation period.

Miso is a power food for a number of reasons. As you often have heard in reference to yogurt, it helps to maintain the intestinal flora. The beneficial lactobacillus microorganisms and enzymes that develop from the fermentation process help digestion and contribute to our ability to resist disease. Miso also has a slightly alkaline effect on the bloodstream. Both of these factors can lessen our susceptibility to radiation.

It is important to realize, however, that only unpasteurized miso is of this quality. Unpasteurized miso is available in bulk in many natural foods stores and by mail order. Heat destroys the living microorganisms, so add miso to food just before serving it. Remove food from the flame and allow a minute or two for flavors to blend; do not cook miso.

Miso is a source of calcium, iron and the B vitamins, nutrients that can block our uptake of radioactive substances. In particular, vitamin B_{12}, which can protect us against cobalt-60, is contained in miso. A recent assessment of the diet of Seventh-day Adventists indicated that the Government-recommended daily amount of 3 micrograms of B_{12} is high. The Seventh-day Adventists live very satisfactorily on a vegetarian diet, which contains smaller quantities of vitamin B_{12}. The general opinion used to be that we need to consume meat protein in order to obtain adequate amounts of B_{12}. However, the experience of the Seventh-day Adventists and other groups clearly indicates otherwise. Maybe the simplified vegetarian diet allows greater absorption of the smaller amount. As a matter of fact, for centuries, Japanese Buddhist monks consumed miso, but no meat whatsoever, and were renowned for their health and longevity.

Miso contains a binding agent called zybicolin, which was identified in 1972, and is effective in detoxifying and eliminating radioactive elements from the body. Research into miso's special properties was prompted by Tatsuichiro Akizuki, a doctor and health researcher from Nagasaki. Akizuki was intrigued by the use of food as preventative medicine, and wrote a book called *Physical Constitution and Food.* Through his studies of both traditional foods native to Japan and modern nutritional science, he formulated a plan that he applied to himself and his co-workers in the St. Francis Clinic. As he used brown rice and miso as the mainstays of his diet, he began to feel better and better. Soon he had everyone in the clinic on a "use food as your medicine" eating plan. He wrote:

> I feel that miso soup is the most essential part of a person's diet. I have found that, with very few exceptions, families which make a practice of serving miso soup daily are almost never sick. I believe that miso belongs to the highest

class of medicines, those which help prevent disease and strengthen the body through continued usage. The basic condition of a person's constitution determines whether or not he will be only mildly or temporarily affected by diseases, or be seriously and chronically affected.

Nagasaki Radiation Survivors Were Protected by Miso

In 1945, when the A-bomb hit not far from the St. Francis clinic, Dr. Akizuki and his staff kept right on working and caring for the flocks of radiation victims. He said:

> I had fed my workers brown rice and miso soup for some time before the bombing. None of them suffered from atomic radiation. I believe this is because they had eaten miso soup. It was thanks to this method that all of us could work away for people day after day, overcoming fatigue or symptoms of atomic disease, and survived the disaster free from severe symptoms of radioactivity.

Confirming this accidental discovery of miso's radioprotective powers, the National Cancer Center of Japan spent thirteen years monitoring people who eat miso soup daily. In 1981, they reported that these individuals are 33 percent less likely to suffer stomach cancer than those who never eat miso soup. Dr. T. Hirayama, who conducted the study, suggested a variety of ways "soybean paste soup" (miso) can be protective: "Some substance in the soy bean or in miso soup may offer resistance to cancer . . . miso provides the body with good nutrition, and therefore makes the body healthier and thus more resistant to cancer . . . the vegetables used in miso soup have some effect in reducing cancer."

With all these benefits, it is lucky that miso also has a delicious taste, aromatic and hearty. For optimal health, you should consume about a teaspoon per day (start by adding one-half teaspoon to your daily diet). In making soup, about half a teaspoon per serving is right. Miso soup is ideal in the morning as it alkalizes and energizes for the day. One or two cups should be taken daily.

There are many other ways to add miso to the daily diet. For example, it makes a flavorful sauce over grains or vegetables, and can be mixed with lemon for salad dressing. Dissolved in hot water, miso makes a satisfying bouillon and a handy substitute for coffee or tea. Travelers can carry it and add it to whatever foods are available. The optimal amount of miso (for travelers and others) is about one-half to one teaspoon a day. One sign of excessive miso consumption is a craving for sweet, sugary foods

Tempeh And Tofu
Note: Caution about the use of Soy foods: avoid G.M.O.

Tempeh and tofu are two traditional protein foods made from soybeans. Thus, they are low on the food chain and carry few contaminants. Both of these soy foods are low in fat and low in calories. They contain the eight essential amino acids, though they have less of the sulfur-containing pair, methionine and cysteine, while being high in lysine. Grains exactly complement this situation. So, soy foods and grains together make a high-quality protein. In terms of a meal, the ideal proportion of grain to soy product is about four to one.

Tempeh. Tempeh is made from cooked soybeans and beneficial bacteria. The action of the bacteria produces a fermented food, similar in texture to chicken and with a flavor like cheese or mushrooms.

Along with miso and sea vegetables (see next page), tempeh is a vegetarian source of vitamin B_{12}, which inhibits the absorption of radioactive cobalt. Tempeh also contains a number of beneficial microorganisms that are typical of fermented foods and check harmful bacteria. Moreover, it contains phytates, which combine with radioactive elements and enable the body to eliminate them.

This traditional food came to America from Indonesia in the 1960s. Since then its popularity has spread. It is sold either fresh or refrigerated in the form of compact eight-ounce cakes less than an inch thick. Tempeh can be sliced and fried or cubed and added to miso soup or a vegetable salad. Since it is already cooked, the cooking time is short, just a few minutes. The meat-like texture allows tempeh to be used in place of ground beef in burgers or chicken in salad.

Tofu. Tofu, the white pillowy squares that you see floating in a vat of water at the Oriental grocery or packaged in the refrigerator at the health food store, is also made from soybeans. In this case, they are mashed into a puree and curdled with the addition of some sea water. The curds are then pressed into squares; the texture is similar to that of a light smooth cheese. (Also like cheese, tofu is rich in calcium.)

A friend who is a stern food critic invited me recently to try a new dish he had created with tofu. Explaining that he'd thought it was a strange food at first, he said—"It's kind of bland, but turns out good."

Actually, tofu *is* bland. But its neutral flavor readily picks up the flavor of spices and vegetables that you cook with, and so it makes a good protein addition to many dishes. Use it with vegetable dishes and rice. Try some cubes floating in soup. It is excellent for sauces and dressings. It is quicker to sauté some tofu than it is to open a can of tomato soup. Both are about the same price per serving, but they are a world apart when it comes to nutrients.

Try to get the freshest tofu you can find. It will keep in the refrigerator about a week if you place it in a bowl of water and change the water every day.

Just experiment with these two soy products and you will be delighted by their versatility!

Sea Vegetables
Note: use caution about sea vegetables since Fukushima.

Who ever heard of eating seaweed? Unless you eat in a Japanese or macrobiotic restaurant, you probably won't come across it. However, the various types of seaweed have been gaining a certain amount of recognition in the U.S. in the last few years. Although you may not be accustomed to seaweeds (except perhaps kelp, which has been popularized lately as a salt substitute), they are a crucial part of a diet for optimal health. This is because they are rich in a variety of minerals and in the radioprotective compound sodium alginate.

In this book I will use the name *sea vegetables,* which emphasizes the importance of seaweeds in the diet—and sounds a lot more appetizing as well!

About 70 percent of the earth's surface is ocean water. The sea was the source of all life forms. It contains every element necessary for life. The components of ocean water are almost identical to those of human blood (only their concentration varies). The minerals and nutrients in seawater are abundantly absorbed in the sea vegetables and are thereby made available to us. Land vegetables can only offer us whatever minerals were obtainable from the soil in which they were grown—and our soil is becoming increasingly depleted from intensive farming methods and modern fertilizers. (In his book on trace elements, Dr. Richard Passwater points out: "Of the 44 minerals and trace elements now found in the sea, 20 are no longer in the land. They are not replaced by the fertilizers which usually contain only nitra phosphates, potassium and occasionally sulfates, or lime which contains calcium.")

Sea vegetables grow in an ocean of minerals and can increase their concentration of minerals such as iodine, calcium, potassium and iron many times. The chart below shows the mineral content of various common sea vegetables. These minerals, along with sodium, are essential to a high state of health and their finest source is sea vegetables. Sea vegetables also contain many trace minerals, including chromium, zinc, magnesium and manganese. Of vitamins, B_{12} is found in good amounts.

A symposium on trace mineral deficiencies was held in Spring 1985 and reviewed in the June 8, 1985 issue of *Science News,* with the comment that trace minerals may play a more important role in human health than had been realized and that their lack can be linked to many problems, such as diabetes and heart disease. The review noted the researchers' warning: "The deficiency levels discussed at the meeting are common in the typical American diet . . . and if such subtle deficiencies are definitely found to jeopardize health they may be affecting millions of people in the U.S. alone."

Mineral Content of Sea Vegetables (mg/100 g)

Sea Vegetable	Calcium	Iron	Iodine	Potassium
Agar-agar	567	6.3	0.2	0
Dulse	296	150.0	8.0	8060
Hijiki	1400	30.0	0	0
Irish moss	885	8.9	0	2844
Kelp	1093	100.0	150.0	5273
Nori, green	470	23.0	0	0
Nori, red	470	23.0	0	0

Sea Vegetables Against Radiation

Note: Use caution with sea vegetables

As we have seen, dietary calcium blocks the human body's incorporation of strontium-90, a similar but radioactive substance, into the bones. One reason why sea vegetables can help to fortify us against the deleterious effects of radiation is their high calcium content. The high amount of calcium in sea vegetables makes them a good way to compensate for the reduction of intake of dairy foods in the diet.

Sea vegetables also contain iodine, which is indispensable to thyroid function. The thyroid influences metabolic efficiency and aids resistance to infection. A deficiency of iodine can result in a lack of energy, an inability to metabolize foods, and weight gain. A slight iodine deficiency could be one factor in the weight problems so prevalent today. For optimal health, we need to satisfy the body's need for iodine on a continuous basis. In addition, experimenters at Johns Hopkins University discovered the impressive fact that when 5 milligrams of iodine are given to an adult for several days, this will reduce the radioactive iodine content in the thyroid gland by about 80 percent.

The compound sodium alginate, contained in sea vegetables, is another radioprotective factor. Sodium alginate binds with radioactive elements and eliminates them from the body. Many studies show this to be true. Such established journals as *Radiation Research, Health Physics, Nature* and the *International Journal of Radiation Biology* are among the many that have published papers on this topic.

In the late 1950s, interest started up. In the introduction to the 1967 book *Strontium Metabolism,* one researcher explained:

> In 1957, while studying the induction of malignant bone tumors by radioactive strontium it occurred to us that inhibition of the carcinogenic action of radioactive strontium might be effected by interfering with the intestinal absorption of the isotope. In the subsequent years, a systematic investigation of substances that might inhibit the uptake of radioactivity from the gastrointestinal tract was undertaken.

The conclusion? "Alginic acid derivatives have been found markedly to inhibit intestinal absorption of radioactive strontium." A study at McGill University in 1964 was one of the first publicized and is often referred to even today. These scientists observed that sea vegetables could reduce by 50 to 80 percent the amount of radioactive strontium absorbed by the intestine. They also found that sodium alginate binds most of the strontium in the intestine while allowing calcium to be absorbed. Dr. Stanley Skoryna, one of the McGill researchers, described this process at the Seventh Annual International Congress of Gastroenterology in Brussels in 1966. He stated:

> The sodium alginate in kelp, a naturally occurring, non-toxic, acidic polysaccharide can discriminate so successfully that it actually separates strontium-90 from the calcium to which it is attached, binding the strontium in the intestine while permitting the calcium to be freely absorbed. The sodium alginate and the bound strontium-90 are subsequently excreted, removing the radioactive material from the body.

Along with Deirdre Waldron-Edward and T. M. Paul, Skoryna also found that besides reducing the absorption, sodium alginate can remove strontium-90 that has already been absorbed: "It is suggested that alginate feeding may be of value in removing previously absorbed ^{90}Sr, introduced by accidental ingestion or by inhalation."

Throughout the 1970s, a series of studies confirming this appeared. Despite such intimidating titles as "Effect of Combined Alginate Treatments on the Distribution and Excretion of an Old Radiostrontium Contamination," the basic ideas are easy to understand. Like fiber and phytates in whole grains and beans, and zybicolin in miso, sodium alginate is a binder or chelating agent.

In theory, the same properties that enable sodium alginate to pick up radioactive substances in your body could cause sea vegetables to pick them up from a contaminated environment. In practice, sea vegetables have been closely monitored and thus far supplies from Japan, the state of Maine and other locations have passed the most rigorous quality control. However, the problem of environmental contamination is something to be aware of for the future.

You can buy sea vegetables in a natural foods store or in an Oriental market. They are dried and usually packaged. Some of the types available are: agar-agar, arame, dulse, hijiki, Irish moss, kombu, nori and wakame. My favorites are *wakame,* which is almost transparent, has a gentle taste and is a plus in any soup; *hijiki,* which is black and stringy and (if not overcooked) makes a splendid side dish; and *kombu,* which is best when placed in the bottom of a cooking pot of rice or beans. It

How Sea Vegetables Protect

Nutrient	Function
vitamin B_{12}	helps build red blood cells; essential for nervous system; blocks uptake of cobalt-60
calcium	strengthens bones; blocks uptake of strontium-90
chromium	helps stabilize blood sugar
iodine	helps maintain thyroid; blocks uptake of iodine-131
iron	builds red blood cells; blocks uptake of plutonium
magnesium	helps maintain pH balance
potassium	helps regulate pH balance of body fluids; blocks uptake of cesium-137
zinc	essential for thymus; blocks uptake of zinc-65
sodium alginate	binds with radioactive and toxic substances

will impart its flavor and make beans more digestible. Kombu is also excellent for making soup stock.

Nuts And Seeds

In Biblical times, nuts and seeds enjoyed a featured place in the diet. In fact, scientists have found that our ancestors were eating nuts way back in prehistory. Nuts and seeds are such unusually substantial foods it is too bad their great potential is passed by in the modern diet. All too often, they are relegated to a minor role, appearing as an occasional snack or as an ingredient in a casserole or salad.

Nuts and seeds are highly concentrated, compact packages prepared by nature to supply all the requirements of a living plant, be it a flower, bush or tree. For this reason, nuts and seeds are endowed with a nearly complete array of vitamins and minerals. They are particularly good sources of B vitamins, vitamin E, calcium, magnesium, potassium, iron and zinc. In addition, nuts and seeds contain almost as much protein as meat, but have the advantage of being lower on the food chain. Foods low on the food chain are less likely to contain high concentrations of chemical or radioactive contamination.

Finally, nuts and seeds are high in essential fatty acids, which we must obtain from foods. These facilitate oxygen transport, assist proteins in building body cells, aid glandular activity and cooperate with vitamin D and calcium. They also help convert carotene into vitamin A in the body.

Nuts and seeds are radioprotective in that they contain fiber (especially pectin) and phytates, binders which facilitate the elimination of radioactive substances. They are also protective because they are such a fine source of nutrients that preclude the uptake of their radioactive counterparts. In addition, the vitamin E in nuts and seeds works as an antioxidant, scavenging for free radicals and helping to repair damage caused by radiation.

Since nuts and seeds are high in fat, they can spoil quickly and become rancid, so be sure to store them in a cool, dry place. Avoid buying nuts that are roasted, salted or processed because their nutritional value is compromised.

Almonds and filberts have the highest calcium content of all nuts and both offer vitamin E and minerals. Almonds are particularly high in magnesium. They are traditionally regarded as having some special healing and protecting properties. One holistic medical doctor I used to work with "prescribed" ten almonds daily for all his patients!

Walnuts, a familiar snack food, offer plentiful iron and potassium. Black walnuts have a higher nutrient content than other varieties. Many walnuts are bleached with lye and sprayed with toxic gases. Be wary and selective when you shop.

Brazil nuts are tough to crack but worth the effort, because they are an excellent source of B vitamins. Pistachios (more fun to shell) are loaded with iron, potassium and thiamine. Pecans provide B vitamins and fatty acids. Peanuts—botanically speaking, a member of the legume family—are rich in vitamin E and niacin and are high in protein. Be careful not to buy peanuts that are rancid or have been contaminated by molds. There are so many "ifs" in commercially manufactured peanut butter that it is hard to vouch for its quality. Thus, it is best avoided.

Sesame seeds, sometimes called the king of seeds, are one of the oldest cultivated crops. They have a high calcium content. Most flavorful when lightly toasted, they can be sprinkled over grains and vegetables. In addition, a delicious and nutritious peanut-butter-like spread called tahini can be made from these seeds. You can also make your own candy by mixing ground sesame seeds with honey.

Pumpkin seeds and sunflower seeds are good sources of pectin. And they are both high in zinc, which is indispensable for the thymus. Sunflower seeds are famous for vitamin B_6, which is needed by the thymus as well. Both sunflower and pumpkin seeds are delicious when freshly toasted in a hot skillet with either a few drops of tamari sauce or some sea salt. You can easily obtain a supply of pumpkin seeds by buying a pumpkin and scraping it out. Then dry the seeds in the oven at a low heat for about an hour. If you grow some of those enormous sunflowers you can also harvest your own sunflower seeds. If you prefer them already grown and hulled, natural food stores usually carry them in bulk.

Nuts and seeds can enrich our diet. There are so many to choose from, you would be nutty not to include them in *your* diet.

A Summary Of The Value Of Various Nutrients

The listing of vitamins, minerals and binding agents that follows will enable you to see the value of various nutrients at a glance. In addition, this section includes a detailed summary of the food sources of selected nutrients. Finally, a Daily Meal Planning Guide and a sample menu are included to help you put the suggestions in this chaper into practice right away.

Vitamins

Vitamin A (carotene) is found in yellow and green vegetables and in beans. An antioxidant, it protects cells by getting rid of free radicals. Vitamin A helps to maintain the skin and mucous membranes. It also helps the body to resist infection—it may even prevent cancer from developing.

The **B-complex** vitamins are-found in whole grains, vegetables, beans, miso, nuts and seeds. The B complex consists of a group of related compounds: vitamins B_a (thiamine,) B_2 (riboflavin), B_6 (pyridoxine) and B_{12} (cyanocobalamin), along with pan-tothenic acid, niacin, biotin and folic acid. The B vitamins all help the immune system and the nervous system. In addition, they play an important role in the metabolism of food.

Vitamin B_6 (pyridoxine) is found in whole grains and vegetables. It is removed from refined grains and not replaced. It is essential for the thymus, which orchestrates the immune system. Among other useful functions, vitamin B_6 also helps build red blood cells.

Vitamin B_{12} (cyanocobalamin) is found in miso, tempeh and sea vegetables. It blocks uptake of cobalt-60 and helps build red blood cells. In addition, it is necessary for proper functioning of the nervous system.

Folic acid is found in green leafy vegetables. It helps build red blood cells.

Vitamin C (ascorbic acid) is found in fresh green vegetables and fruit. It supports the immune system, is necessary for the adrenal glands, helps build blood and assists in the utilization of calcium. It is an antioxidant and it counteracts pollutants.

Vitamin E (tocopherol) is found in grains, vegetables, nuts, seeds and oils. It, like vitamin C, is versatile. Vitamin E aids the heart and circulation and counteracts free radicals.

Minerals

Calcium is found in green vegetables, sesame seeds, almonds, beans, sea vegetables, whole grains, tofu and miso. Ninety-nine percent of the calcium in the body is in the structural system. The rest regulates nerves and muscles and blood pH. Dietary calcium can help block the body's uptake of radioactive strontium-90.

Iodine is found in sea vegetables and some land vegetables (if they are grown on soil containing iodine). It is necessary for the function of the thyroid gland, and blocks uptake of radioactive iodine-131.

Iron is found in leafy green vegetables, whole grains, sea vegetables, beans, nuts and seeds. It is essential for building red blood cells. In addition, it blocks uptake of plutonium.

Potassium is found in all vegetables and beans. It is present in sea vegetables as well. Potassium helps regulate the pH balance of body fluids and blocks uptake of cesium-137.

Chromium is found in whole grains and sea vegetables. It is essential in maintaining the blood sugar level. (Problems here can contribute to the development of hypoglycemia or diabetes.)

Magnesium is found in green leafy vegetables, whole grains, sea vegetables and nuts (especially almonds). It helps maintain the body's pH balance and has a calming effect on the nerves.

Selenium is found in whole grains, garlic and some vegetables. Some agricultural areas have very small amounts available from the soil. Selenium counteracts free radicals, thus protecting us against radiation and stimulating the immune response.

Zinc is found in whole grains and green vegetables (if they are grown on soil containing zinc), sea vegetables, nuts and seeds. This mineral is essential for the thymus and blocks the body's uptake of zinc-65. It is also involved in maintaining the blood sugar level, growth and healing, and mental equilibrium.

Factors that Bind Radioactive Substances

Fiber. Found in whole grains, vegetables, nuts, seeds and beans. Pectin is a type of fiber found primarily in seeds and fruit.

Phytates. Found in grains and beans.

Sodium alginate. Found in sea vegetables.

Sulfur-containing amino acids. Found in some vegetables, especially those in the cabbage family. In addition to binding radioactive substances, sulfur-containing vegetables may also contain compounds that inhibit the development of cancer.

Zybicolin. Found in miso.

NOTE: Since the cumulation of radiation releases from the 400-some nuclear power plants world-wide and the accidents small and large it is suggested to get a greenhouse, or protected garden, with filtered water (avoid rainwater). Otherwise do the best you can following the guidelines in this book. Read the website: www. nuclearreader.info for a concise overview of our situation.

VITAMIN A

Food Source	Per Serving (100 g/3½ oz)*
VEGETABLES	
carrots	11,000 IU
leafy greens	7,000 to 10,000
squash	7,000
broccoli	3,300
brussels sprouts	676
BEANS	
chick peas	50
soybeans	30
SEA VEGETABLES	
nori	20,000
hijiki	150
wakame	140
FRUIT	
cantaloupe	3,400
apricots	2,700
peaches	1,330
watermelon	590
apples	117
grapes	100

Daily Intake

DIET
1 leafy green
1 yellow vege-
table

SUPPLEMENT
not necessary

RDA
5,000 IU

Protective Functions

protects from free radicals; supports immune functioning; maintains skin and mucous membranes

Key To Units Used

g, grams; IU, international units; meg, micrograms; mg, milligrams; oz, ounces

* unless otherwise specified

VITAMIN B$_6$

Food Source	Per Serving (100 g/3½ oz)*	Daily Intake
		DIET
GRAINS		1 grain
brown rice	0.4 mg	1 bean
		2 vegetables
VEGETABLES		
		SUPPLEMENT
root vegetables	0.2 to 0.3	2 mg or more
leafy greens	0.2	
		RDA
BEANS		2.2 mg
soybeans	0.8	
lentils	0.6	**Protective Functions**
FISH		essential for
salmon	0.7	thymus to make T-cells; helps build red blood
OTHER		cells; combats
sirloin steak	0.9 (8 oz)	stress

Key To Units Used

g, grams; IU, international units; meg, micrograms; mg, milligrams; oz, ounces

* unless otherwise specified

VITAMIN B$_{12}$

Food Source	Per Serving (100 g/3½ oz)*	Daily Intake
SEA VEGETABLES		**DIET**
all varieties	0.3 to 6.0 mcg	1 sea vegetable
		1 serving
SOY PRODUCTS		miso
		plus occasional
miso	0.03 per tablespoon	tempeh and fish
tempeh	0.03 per tablespoon	
		SUPPLEMENT
FISH		not necessary
common varieties	0.2 to 0.9	**RDA**
		3.0 mcg

Protective Functions

blocks uptake of cobalt-60; helps build red blood blood cells

Key To Units Used

g, grams; IU, international units; meg, micrograms; mg, milligrams; oz, ounces

* unless otherwise specified

VITAMIN C

Food Source	Per Serving (100 g/3½ oz)*
VEGETABLES	
leafy greens	150 mg (appx.)
broccoli	113
brussels sprouts	100
cauliflower	78
cabbage	47
FRUIT	
strawberries	59
oranges	50 (per orange)
cantaloupe	33
nectarines	13
apricots	10

Daily Intake

DIET
2 to 3 vegetables fruit in season (especially from your area)

SUPPLEMENT
depending on pollution and stress, 0 to 1,000 mg

RDA
60 mg

Protective Functions

protects from free radicals; supports immune system; detoxifies pollution; helps build blood

Key To Units Used

g, grams; IU, international units; meg, micrograms; mg, milligrams; oz, ounces

* unless otherwise specified

VITAMIN E

Food Source	Per Serving (100 g/3½ oz)*	Daily Intake
GRAINS		*DIET* 3 vegetables
oatmeal	3 to 4 IU	3 grains
all whole grains	3 to 4	1 teaspoon oil
VEGETABLES		*SUPPLEMENT* 50 to 400 IU d-alpha
leafy greens	1 to 4	tocopherol
turnips	1 to 4	
		RDA
BEANS		30 IU
all varieties	1 to 4	**Protective Functions**
VEGETABLE OILS		
unrefined	12 to 20 per tablespoon	protects from free radicals; aids heart and circulation

Key To Units Used

g, grams; IU, international units; meg, micrograms; mg, milligrams; oz, ounces

* unless otherwise specified

CALCIUM

Food Source	Per Serving (100 g/3½ oz)*
VEGETABLES	
broccoli	100 mg
leafy greens	90 to 200
cabbage	67
green beans	56
onions	51
turnips	39
BEANS	
soybeans	226
chick peas	150
kidney beans	130
SOY PRODUCTS	
tofu	128 to 139
tempeh	70 to 105
miso	68
SEA VEGETABLES	
hijiki	1,400
wakame	1,300
kombu	800
nori	470
SEEDS AND NUTS	
sesame seeds	1,160
almonds	282
brazil nuts	186
sunflower seeds	140
FISH	
common varieties	20 to 90
OTHER	
milk	118

Daily Intake

DIET
1 leafy green
1 vegetable
1 sea vege-
   table
1 bean

SUPPLEMENT
not necessary

RDA
800 to 1,200 mg

Protective Functions

blocks uptake of strontium-90; helps maintain blood pH; strengthens bones

Key To Units Used

g, grams; IU, international units; meg, micrograms; mg, milligrams; oz, ounces

* unless otherwise specified

IODINE

Food Source	Per Serving (100 g/3½ oz)*	Daily Intake
SEA VEGETABLES		**DIET** 2 servings sea vegetables
arame	90 to 560 mcg	
kombu	193 to 470	**SUPPLEMENT** in event of increased radiation exposure,† 150 mg potassium iodide in tablet form
hijiki	40	
wakame	18 to 35	
FISH		
shellfish	0.29	**RDA** 150 mcg
common varieties	0.07	

Protective Functions

helps maintain thyroid; blocks up-take of iodine-131

Key To Units Used

g, grams; IU, international units; meg, micrograms; mg, milligrams; oz, ounces

* unless otherwise specified

+ Check with a qualified health care provider under these circumstances.

IRON

Food Source	Per Serving (100 g/3½ oz)*	Daily Intake
GRAINS		**DIET**
millet	6.8 mg	2 grains
oats	4.6	1 leafy green
whole wheat	3.0	1 bean
		1 sea vegetable
VEGETABLES		
leafy greens	1.7 to 3.3	**SUPPLEMENT** not necessary
BEANS		**RDA**
chick peas	6.9	10 to 18 mg
lentils	6.8	
aduki beans	4.8	
SEA VEGETABLES		**Protective Functions**
hijiki	29.0	
nori	23.0	blocks uptake of plutonium; builds red blood cells
kombu	15.0	
wakame	13.0	
SEEDS		
sesame seeds	10.0	
sunflower seeds	7.1	
FISH		
sardines	2.9	
OTHER		
sirloin steak	7.2 (8 oz)	

Key To Units Used

g, grams; IU, international units; meg, micrograms; mg, milligrams; oz, ounces

* unless otherwise specified

SELENIUM

Food Source	Per Serving (100 g/3½ oz)*		Daily Intake
GRAINS			**DIET** 2 or 3 grains
brown rice	40 mcg		
barley	30 to 40		**SUPPLEMENT** 100 to 200 mcg
pasta	0 to 60		
oats	3 to 11		**SUGGESTED AMOUNT** 50 to 200 mcg
VEGETABLES			
broccoli	amounts vary; some		**Protective Functions**
garlic	soil is depleted		
FISH			protects from free radicals; reinforces immune system
all varieties	15 to 40		

Key To Units Used

g, grams; IU, international units; meg, micrograms; mg, milligrams; oz, ounces

* unless otherwise specified

SULFUR

Food Source	Per Serving (100 g/3½ oz)*	Daily Intake
GRAINS		DIET 1 or 2 vege- tables
brown rice	10 mg	
barley	240	SUPPLEMENT 500 to 2,000 mg
VEGETABLES		SUGGESTED AMOUNT 850 mg
kale	8,600	
watercress	5,390	
brussels sprouts	3,530	
cabbage	1,710	Protective Functions
turnips	1,210	
cauliflower	1,186	
onions	265	resists radiation on cellular level; coun- teracts toxins; blocks
BEANS		uptake of sulfur-35; helps repair DNA
lima beans	260	
lentils	120	
OTHER		
eggs	67 (per egg)	

Key To Units Used

g, grams; IU, international units; meg, micrograms; mg, milligrams; oz, ounces

* unless otherwise specified

ZINC

Food Source	Per Serving (100 g/3½ oz)*
GRAINS	
oatmeal	14.0
corn	2.5
brown rice	1.5 mg
VEGETABLES	
green peas	1.6
parsley	0.9
cabbage	0.8
carrots	0.5
watercress	0.5
BEANS	
split peas	4.2
black beans	0.4
lentils	0.2
NUTS	
pecans	4.5
walnuts	3.6
FISH	
haddock	1.7
shrimp	1.5

Daily Intake

DIET
often is not adequate; avoid frozen vegetables

SUPPLEMENT
15 to 50 mg

RDA
15 mg

Protective Functions

essential for thymus to make T-cells; blocks uptake of zinc-65

Key To Units Used

g, grams; IU, international units; meg, micrograms; mg, milligrams; oz, ounces

* unless otherwise specified

FIBER

Food Source	Per Serving (100 g/3½ oz)*
GRAINS	
rolled oats	2.8 g
barley	2.2
cornbread	1.4 (1 piece)
brown rice	1.3
spaghetti	0.8
VEGETABLES	
corn	3.9
broccoli	2.6
leafy greens	2.5
squash	2.2
cabbage	2.1
radishes	2.0
brussels sprouts	1.8
carrots	1.0
BEANS	
lentils	3.7
kidney beans	2.2
NUTS	
almonds	5.1
brazil nuts	1.5
pistachio nuts	1.0
FRUIT	
raspberries	4.0
apples	3.9
blueberries	2.5
peaches	1.4

Daily Intake

DIET
3 grains
4 vegetables
1 bean

SUPPLEMENT
not necessary

RDA
10 g

Protective Functions

binds with radioactive and toxic substances

Key To Units Used

g, grams; IU, international units; meg, micrograms; mg, milligrams; oz, ounces

* unless otherwise specified

Daily Meal Planning Guide

The easiest way to incorporate protective foods into your daily intake is probably to separate them into groups and to remember the number of servings per group. You can make a list and tack it up in the kitchen. Consult this list often, and very soon this way of eating will be second nature to you! If you generally follow this plan, you will satisfy all your daily nutrient needs and also become your own anti-pollution device.

Meal Planning Guide

Food Group	Ratio	Servings
Grains whole grains, noodles and pasta	35-45%	4 or 5
Vegetables leafy greens cabbage family yellow vegetables	20-30%	2 or 3 2 or 3 1 or 2
BEANS lentils, chick peas, etc. bean products, tempeh, tofu	5%	1 or more 1
SEA VEGETABLES (any type) in soup (any type) in side dishes	5%	1 1
SEEDS AND NUTS any type(s)	less than 5%	1
FISH small white-meat fish	less than 5%	a few times per week
FRUIT in season	less than 5%	

Note: Caution on sea vegetables and soy foods. Consider reducing grains and increasing vegetables

Sample Menu

A daily meal plan might look something like this:

Meal	Number of Servings	Food Group
BREAKFAST		
miso-wakame soup	1	soups (with sea vegetable)
oatmeal	1	grains
SNACK		
apple	1	fruit
LUNCH		
noodles with sesame		
sauce	2	grains
steamed greens	2	vegetables (green leafy)
SNACK		
almonds	1	seeds and nuts
DINNER		
fried rice with cabbage	1	grains
and chick peas	1	vegetables (cabbage family)
	1	beans
steamed broccoli	1	vegetables (cabbage family)
ginger, carrots and	1	vegetables (yellow)
hijiki	1	sea vegetables

Total Servings per Person

Grains	4	Sea Vegetables	2
Leafy Greens	2	Soups	1
Cabbage Family	2	Seeds and Nuts	1
Yellow Vegetables	1	Fruit	1
Beans	1		

FOOD IRRADIATION

Something very serious and very undermining to the possibilities of utilizing food for its protective powers has recently surfaced. This is the plan to *irradiate* food. Food irradiation is a process of food preservation technology that is aimed at killing certain bacteria and extending the shelf life of foods. This silent, invisible threat to our food supply is not in the far-off future. It is already endangering us. Food irradiation was approved by the U.S. Food and Drug Administration on April 18, 1986 for use on *fruits, vegetables* and *grains.* It had already been implemented for use on spices; the April ruling tripled the allowed dose. In addition, other regulations have cleared the way for the irradiation of pork.

Food irradiation uses gamma rays from cesium-137 and cobalt-60. Like other forms of ionizing radiation, this exposure causes chemical changes. The substances that are formed in foods exposed to irradiation are referred to as *radiolytic products. A* number of these radiolytic products, which do not occur naturally in the irradiated foods, have been identified. They include harmful free radicals such as hydrogen peroxide and potential cancer-causing and mutation-causing chemicals such as benzene and formaldehyde.

Moreover, food irradiation affects the nutrients in foods. It depletes the antioxidant vitamins A, C and E, destroys some of the B vitamins and alters proteins and fats.

Proponents of food irradiation claim that the changes in the chemical and nutritional composition of irradiated foods are harmless. Some studies, however, have suggested that the consumption of irradiated foods may cause cancer, kidney disease, liver disease, birth defects and blood abnormalities. *The fact is that no long-term studies have been done!*

In addition to potential health hazards to consumers, the radiation used to preserve foods poses a danger to workers in the processing plants. Like the operation of a nuclear power plant, food irradiation must be done with care. It involves doses of 100 kilorads (100,000 rads)—about 200 times the lethal dose for human beings. There is also danger to the community at large from the use, long-distance transport and disposal of radioactive materials. Cesium-137 and cobalt-60 weren't chosen for use at random. They are byproducts of the nuclear power industry. Each food irradiation plant will require millions of curies of cesium. As you may recall from our discussion of the food chain, cesium-137 is a particularly threatening environmental contaminant, as plants and animals can absorb and use it in place of the essential nutrient potassium.

It is of the utmost importance that we join the fight against food irradiation today and let our legislators know of our opposition to this practice. In the meantime we can protect ourselves by avoiding fast, junk, frozen and packaged foods that may contain hidden irradiated ingredients, and by purchasing natural whole foods instead. (Legally, these would have to be labeled if irradiated.) In addition, we may consider the use of protective supplements.

7

Supportive Supplements

I have heard the nineteen-sixties described as the decade of vitamins, the seventies the decade of minerals and the eighties the decade of amino acids. It is true the mega vitamin pioneers, including Osmond, Hoffer, Pauling and Pfeiffer, appeared in the sixties. They lauded vitamins as the cure-all for many ailments. More recently, minerals have had their day in the sun, with zinc recognized in relation to schizophrenia and chromium in relation to hypoglycemia and diabetes. Now, amino acids with strange-sounding names like *phenylalanine* and *tryptophan* are being utilized against depression and insomnia.

But shouldn't we get all the vitamins, minerals and amino acids we need from a balanced diet? It seems we should; it seems nature should know best. And frankly, I think nature *did* know best. But two things have changed in this century: the food that nature wanted to provide, and the environment that nature wanted to provide.

First of all, our overused farmland does not have rich soil. Crops grown there cannot contain nutrients that are not available to them. Moreover, food refining methods take out twenty-one essential vitamins and minerals—and replace only B_1, B_2, niacin and iron. Second, our environment is full of pollutants that eat up our nutrient supplies. The unprecedented stresses of modern living adversely affect the body's calcium reserves and also reduce our ability to absorb vitamin C and the B vitamins. In light of these facts of life, it becomes apparent that we do not have sufficient nutritional reserves to do the life-preserving work of detoxifying the pollution and low-level radiation with which we inevitably come into contact.

Many studies confirm the unfortunate truth that environmental factors can threaten even the best-nourished individuals among us. Thus, the judicious use of supportive supplements is often a good idea. The purpose of this chapter is to familiarize you with the supplements that are most effective in radioprotection. On the pages that follow, we will discuss supplements for the immune system, detoxifying supplements, and supplements for fortification and general overall benefits. A chart of suggested supplements is also included.

Supplements For The Immune System

Like many nutrients, vitamin B_6 and zinc have a multitude of functions in the human body. They are worthy of individual consideration, however, because of all the supplements they relate most directly to the immune system, which is the body's front line of defense.

Vitamin B_6

Vitamin B_6, which is also known as pyridoxine, is a member of the B-complex group. The B-complex vitamins work cooperatively to maintain metabolism, while each separate B vitamin also does its own particular job. It has been said that B_6 deficiency is the most prevalent vitamin deficiency in the United States today. If you eat a lot of meat protein, fats and refined foods, you may have a deficiency of vitamin B_6.

Vitamin B_6 has the crucial task of nourishing the thymus gland (see page 84) so it will produce healthy T-cells. The T-cells destroy malformed cells that can be caused by the presence of free radicals. In addition, B_6 is an important factor in the manufacture of healthy red blood cells, is helpful in preventing kidney stones and is necessary for maintaining mental equilibrium and the nervous system.

There is also some evidence of a relation between high dietary fat intake and vitamin B_6 needs. When the fat intake is higher, the needs for B_6 are higher. And lack of B_6 can cause accumulation of fats in the liver. Another factor influencing the body's need for vitamin B_6 is the oral contraceptive pill. Women taking the pill need to have larger amounts of B_6.

Lack of B_6 may predispose people to atherosclerosis (thickening and hardening of the arteries). As it is needed for the proper metabolism of protein, if vitamin B_6 is not available, the toxic byproducts of poorly metabolized proteins may be deposited on the walls of the arteries. This is the beginning of the formation of plaques that decrease the diameter of the blood vessels, making it harder for them to carry the volume of blood. High blood pressure can develop as a result. Perhaps vitamin B_6 depletion is one reason why women taking the pill have a death rate from cardiovascular disease up to ten times greater than that of women who do not take the pill.

Plaques usually form behind small bumps (benign tumors) in the blood vessels. These tumors may be related to alpha radiation exposure from tobacco smoke or other environmental contaminants. Thus, smoking (active and passive) can increase the risk of cardiovascular disease.

Zinc

Zinc has been known to play a role in human metabolism since the nineteen-thirties, but it wasn't until the sixties that zinc deficiency was understood to have many serious effects. Zinc was first included on the list of essential nutrients by the National Academy of Sciences in 1974. Along with vitamin B_6, zinc is essential for the optimal functioning of the thymus gland, which plays an important role in immunity. In its natural, stable form, this mineral can block the body's uptake of radioactive zinc-65. Zinc is also a constituent of enzymes involved in protein synthesis and carbohydrate metabolism. It is involved in the production of insulin and nucleic acids (DNA and RNA) as well. Moreover, zinc influences mental equilibrium and the body's rate of healing. People who take supplemental zinc before an operation are often surprised at the speedy healing.

The amount of zinc available in foods depends on the amount available in the soil. If plants can get enough zinc from the soil, then we can get enough zinc from plants—but only if we eat them in an unprocessed form. Eighty percent of the zinc is removed from wheat flour when it is refined. Frozen green vegetables have most of their zinc removed in order to preserve their bright green color.

Dr. Carl Pfeiffer, an expert on zinc metabolism, believes the whole human population is bordering on zinc deficiency and suggests that we take supplements in tablet form. A recent analysis of institutional diets showed that they provided about 8 to 11 milligrams of zinc—and in some cases less. The United States RDA for adults, however, is 15 milligrams.

In relation to other minerals, sufficient intake of zinc will help the body eliminate toxic lead and cadmium. It also displaces copper, which is significant because high amounts of copper have been linked with depression of the immune system. Unfortunately, however, several factors can deplete our body's reserves of zinc. Oral contraceptives can have this effect. In addition, vitamin pills that contain as little as 2 milligrams of copper can overpower the benefits we gain from zinc (copper is an antagonist to zinc). Water supplied through copper pipes may also deplete our zinc reserves.

In fact, low zinc and high copper is not an unusual mineral imbalance these days, because of soil depletion and copper pipes. One of my nutrition clients had a very high level of copper. Reviewing her dietary habits, I discovered that she drank a lot of ice water—which came through a copper pipe in her refrigerator! To avoid zinc deficiency, we must be careful to avoid sources of excess copper.

Fiber and phytic acid, which are found in whole grains and beans, are valuable protectors against the negative health consequences of radiation. However, these food substances interfere with our absorption of zinc. So, even when eating as we propose, it is wise to take a supplement (make sure, however, that the supplement you choose does not contain copper).

One sign of zinc deficiency is reduced sensitivity to the taste of foods. The zinc-deficient are always saying "You forgot to salt the food." They tend to use excessive salt for this reason. Another sign, in children, is slow growth rate.

Two Special Supplements

There are two especially effective detoxifying supplements that are worthy of attention. They are *cysteine,* an amino acid, and garlic tablets made from concentrated, deodorized garlic.

Cysteine

The cabbage family of vegetables contains the sulfur amino acid cysteine, which has the ability to scavenge free radicals, counteracting the damage of low-level radiation. Cysteine can also detoxify pollutants, such as the heavy metals lead, mercury and cadmium, by binding with them.

In 1978 an article entitled "Mechanisms of Radioprotection—A Review" outlined the findings of researchers at the National Institutes of Health in Bethesda, Maryland. In addition to the action of undoing free radical formation, author Edmund Cope-land reviews three other mechanisms at the molecular level, including the alteration of cell membranes by sulfur so they become temporarily radio-resistant. He concludes that each radioprotective mechanism "has deficiencies but the actual phenomena of chemical protection probably involve elements of each."

In general, two or three daily servings of cabbage-family vegetables offer protection. A supplement of L-cysteine gives insurance. Several different brands are on the market; 500 to 1,000 milligrams per day, or up to 2,000 milligrams per day on a temporary basis (specifically, if you have been exposed to excess radioactivity), is recommended. This amount should be balanced with three times as much vitamin C and ample vitamin B_6.

Garlic Tablets

Garlic, a member of the lily family, is among the oldest cultivated plants. Its medicinal properties have long been recognized all over the world. An Egyptian papyrus known as the *Codex Ebers,* dating to about 1,500 BC, details about eight hundred healing remedies. The codex includes twenty-two remedies employing garlic for a variety of ailments: heart problems, tumors, headaches, bites and worms.

Students of ancient history have also discovered references to garlic being used by athletes in the Olympic Games in Greece. Throughout the years, folk medicine has called for its use as an antiseptic, sort of an herbal antibiotic. Louis Pasteur confirmed the antibacterial properties of garlic in 1860.

In addition, recent research shows that garlic can lower blood pressure and dissolve blood clots. There is also evidence that it detoxifies heavy metals and low-level radiation. Garlic contains selenium, which fights free radicals. The pungent oil in garlic is *allicin,* a sulfur compound that is a recognized antimicrobial agent. Thus, garlic fortifies our immune system and overall health in a variety of ways.

Unfortunately, allicin is unstable. To preserve its potency, garlic must be aged and treated in a special way—without any heat or acid. In my nutrition practice, I have found that the garlic supplement called Kyolic,® which is made from such specially treated garlic and cured for nearly two years, is effective in detoxifying heavy metals.

A good way to derive the maximum benefit from Kyolic® is to take four tablets a day but only five days a week, or every day for three weeks out of four. The breaks are important as they give the body a rest. In case of extra exposure to contaminants, take up to six tablets a day. Each tablet contains 300 milligrams of garlic extract. To eat a comparable amount of garlic would probably cause a number of side effects.

A Handy Reference Guide To Selected Supplements

This list of selected supplements describes six vitamins and eight minerals according to a variety of characteristics: their function, what conditions they are therapeutic for, what other factors work synergistically with them, what factors limit their effectiveness, what deficiency symptoms can occur in their absence, what amounts to consider using, and what foods are good sources.

The amounts recommended for each vitamin and mineral are listed three ways: according to the Recommended Daily Allowances (where applicable), according to suggestions for radio-protection, and according to what might be appropriate in an emergency—i.e., in case of extraordinary exposure to low-level radiation. Each of these amounts is suggested for daily intake, unless otherwise indicated.

In some cases no supplements are suggested. In all cases there are many variables—your state of health, your food intake and the current state of the environment—that must be taken into consideration. The general guidelines are parameters to outline a range of choice. For example, for persons living within a certain radius of Chernobyl after the accident in April 1986 or subject to its distant windswept fallout, it would be recommended to ensure that ample quantities of the nutrients which block radiation were obtained and that the intake of nutrients which fortify immunity and scavenge free radicals be increased. But each case and situation have so much individuality, that an absolute number is not appropriate. Individuals need to exercise their best judgment in utilizing the suggestions given here.

The radio-protective effect of some plants has been documented in published research. Among these to consider including in your intake are: holy basil (ocimum sanctum), siberian ginseng (Eleuthero), ashwaganda, dandelion root, milk thistle, turmeric (curcumin), rose hips, rosemary, parsley, garlic, and onion. It is suggested to grow your own and make a tea or tincture.

A Guide to Selected Supplements

Vitamin A

Functions

Bolsters resistance to infection; maintains epithelial tissues; aids vision; plays a role in growth and repair of tissues; helps form bones and teeth; is a factor in the prevention of cancer; protects from toxins; functions as antioxidant.

Therapeutic For

Infections; allergies; skin problems.

Amounts

RDA: 5,000 IU men and women.
Protective: 10,000 IU.
Emergency: 10,000–25,000 IU.

Supplements

Best utilized form is beta-carotene; take protective dosage daily if desired.

Limiting Factors

Nitrites; antacids; alcohol; caffeine; cortisone; too much iron; vitamin D deficiency.

Synergists

B-complex vitamins; vitamins C and E; zinc.

Deficiency Symptoms

Poor night vision; dry skin; susceptibility to infection.

Signs of Overdosage

Yellowing of skin; enlarged liver and spleen; loss of hair; skin peeling.

Food Sources

Green and yellow vegetables, especially leafy greens, broccoli, spinach, watercress, squash and carrots; apricots.

Vitamin B$_6$

Functions

Necessary for thymus gland to produce T-cells; helps form red blood cells; combats stress.

Therapeutic For

Dermatitis; seizures; water retention; nausea of pregnancy; mental disturbances; anemia.

Amounts

RDA: 2.2 mg men, 2.0 mg women.
Protective: 10 mg (best obtained from foods).
Emergency: for extra support to thymus, take low-dose supplement, 10–100 mg.

Supplements

Vitamin B$_6$ should be adequate in a diet with sufficient grains and vegetables, but need is contingent on presence of limiting factors.

Limiting Factors

Oral contraceptives; stress; alcohol; antibiotics; diets high in refined foods.

Synergists

B-complex vitamins, especially pantothenic acid; zinc.

Deficiency Symptoms

Infections; weakness; lack of dream recall.

Food Sources

Whole grains; beans; vegetables.

Vitamin B$_{12}$

Functions

Essential for red blood cells; prevents anemia; helps nervous system; blocks uptake of cobalt-60.

Therapeutic For

Anemia; fatigue; alcoholism; stress.

Supplements

Depending on overall diet, occasional B$_{12}$ pill may be appropriate.

Limiting Factors

Oral contraceptives; stress; alcohol; antibiotics.

Synergists

B-complex vitamins, especially

Amounts

RDA: 3.0 mcg men and women.
Protective: 2,000 mcg sublingual (under the tongue).
Emergency: 2,000 mcg sublingual (under the tongue).

vitamin B$_6$ and folic acid; vitamin C.

Deficiency Symptoms

Anemia.

Food Sources

Tempeh; miso; sea vegetables.

Pantothenic Acid

Functions

Maintains the thymus gland; aids metabolism; counteracts stress; nourishes the adrenal gland.

Therapeutic For

Loss of hair; premature aging; prematurely gray hair; eczema.

Amounts ·

RDA: none stated; many authorities consider optimal daily intake to be 5 to 10 mg.
Protective: Best obtained from foods.
Emergency: Low-dose B-complex supplement.

Supplements

Not necessary.

Limiting Factors

Alcohol; caffeine; stress.

Synergists

All B-complex vitamins; vitamin C.

Deficiency Symptoms

Furrowed tongue.

Food Sources

Whole grains; all vegetables.

Vitamin C

Functions

Maintains the adrenal gland and connective tissue; helps the body to utilize calcium; supports the immune system; functions as antioxidant; detoxifies pollution; helps build blood.

Therapeutic For

Arthritis; backaches; some schizophrenias.

Amounts

RDA: 60 mg men and women.
Protective: 500–3,000 mg.
Emergency: 3,000–10,000 mg.

Supplements

Considering stress and pollution, it is generally good to take between 500 mg and 3,000 mg daily.

Limiting Factors

Stress; smoking; alcohol; oral contraceptives; antibiotics; carbon monoxide; pollution.

Synergists

All other vitamins.

Deficiency Symptoms

Low resistance; gum problems; weak capillaries; tendency to bleed or bruise easily.

Food Sources

Fresh green vegetables; fruit.

Vitamin E

Functions

Helps keep heart healthy; builds blood; is an anticoagulant and antioxidant.

Therapeutic For

High cholesterol levels; poor blood flow; symptoms of aging; exposure to pollution.

Limiting Factors

Mineral oil; iron; chlorine.

Synergists

Vitamin A; B-complex vitamins.

Deficiency Symptoms

Heart disease; aging symptoms; anemia; circulation problems.

Amounts

RDA: 30 IU men and women.
Protective: 50–400 IU.
Emergency: 200–600 IU.

Supplements

Protective dose of 50–400 IU.

Food Sources

Leafy green vegetables; unrefined vegetable oils; whole grains; nuts; seeds.

Calcium

Functions

Strengthens bones; essential for nervous system; buffers acids and maintains pH; combats stress to protect immune system; blocks uptake of strontium-90.

Therapeutic For

Brittle bones; tension.

Amounts

RDA: 800 mg men and women (some authorities believe that the RDA should be somewhat higher).
Protective: Leafy greens and sea vegetables.
Emergency: If leafy greens and sea vegetables are not available to fulfill daily requirement, any calcium supplement *except* bone meal may be used.

Supplements

Not necessary if diet is adequate.

Limiting Factors

Stress; lack of magnesium; caffeine; lack of vitamin D.

Synergists

Phosphorus; magnesium; vitamin C; vitamin A; vitamin D from sunlight; exercise.

Deficiency Symptoms

Nervousness; insomnia; bone loss.

Food Sources

Green vegetables; beans; whole grains; sea vegetables; nuts and seeds.

Chromium

Functions

Maintains blood sugar level by ensuring proper use of insulin, and so helps carbohydrate metabolism.

Therapeutic For

Diabetes, poor blood circulation.

Amounts

RDA: none stated, but essential.
Protective: 100–300 mcg.
Emergency: 100–300 mcg.

Supplements

If diet is or has been high in refined foods, 100–300 mcg.

Limiting Factors

Diets high in refined foods.

Synergists

Zinc; manganese; magnesium.

Deficiency Symptoms

Diabetes; hypoglycemia; atherosclerosis.

Food Sources

Whole grains; sea vegetables.

Iodine

Functions

Necessary for thyroid gland, which regulates metabolism; blocks uptake of iodine-131.

Therapeutic For

Obesity; fatigue; low blood pressure.

Amounts

RDA: 150 mcg men and women.
Protective: best from food sources.
Emergency: 150 mg potassium iodide in tablet form.

Supplements

Not necessary if dietary intake is adequate.

Limiting Factors

Damaged thyroid that cannot utilize this nutrient.

Deficiency Symptoms

Goiter (enlarged thyroid); low energy; dry skin and hair; easy weight gain.

Food Sources

Sea vegetables; fish; seafood.

Iron

Functions

Component of hemoglobin; increases body's resistance; blocks uptake of plutonium.

Therapeutic For

Anemia.

Amounts

RDA: 10 mg men, 18 mg women.
Protective: best from food sources.
Emergency: if leafy greens and other food sources are contaminated, take 10–18 mg in supplement form.

Supplements

Not necessary if dietary intake is adequate.

Synergists

Vitamin B_6; vitamin B_{12}; folic acid; vitamin C; magnesium.

Limiting Factors

Caffeine; excess phosphorus.

Deficiency Symptoms

Anemia; low energy; pallid complexion; brittle fingernails.

Food Sources

Leafy greens and sea vegetables; beans; whole grains; nuts and seeds.

Magnesium

Functions

Aids metabolism of carbohydrates and proteins; helps maintain pH; calmative.

Therapeutic For

Obesity; alcoholism; nervous system.

Amounts

RDA: 350 mg men, 300 mg women.

Limiting Factors

Physical exertion.

Deficiency Symptoms

Inappropriate nervousness.

Protective: best from food sources.

Emergency: if you are taking a calcium supplement, take a magnesium supplement that equals half the dose of the calcium.

Supplements

Not necessary.

Synergists

Calcium; vitamin B$_6$; vitamin C.

Food Sources

Green leafy vegetables; sea vegetables; whole grains; nuts and seeds.

Potassium

Functions

Manages with sodium the transport of nutrients across cell membranes; helps regulate pH balance; blocks uptake of cesium-137; helps control activity of heart, kidneys and skeletal muscles; aids the liver; called "the healer."

Therapeutic For

Weakness; nervousness; high blood pressure.

Amounts

RDA: none stated; many authorities consider optimum daily intake to be about 1,000 mg.
Protective: best source is vegetables.

Emergency: best from food sources.

Supplements

Not necessary.

Synergists

Sodium; vitamin B_6.

Limiting Factors

Caffeine, diuretics.

Deficiency Symptoms

Muscle cramps; irregular heart beat; high blood pressure.

Food Sources

Vegetables; beans; sea vegetables.

Selenium

Functions

Stimulates immune response; functions as antioxidant; detoxifies pollution and combats its effects.

Amounts

RDA: none stated; authorities suggest 50–200 mcg.
Protective: 100–200 mcg.
Emergency: 200 mcg.

Supplements

100–200 mcg, preferably from a non-yeast source.

Synergists

Vitamin A, vitamin E.

Therapeutic For

Toxicity.

Limiting Factors

Some soils are deficient in selenium.

Deficiency Symptoms

Premature aging; arthritis; dandruff.

Food Sources

Whole grains; vegetables; garlic; fish.

Zinc

Functions

Aids growth and tissue repair; normalizes blood sugar; essential for the thymus to make T-cells; promotes mental equilibrium; blocks uptake of zinc-65.

Therapeutic For

Certain types of schizophrenia; alcoholism; hyperactive children; Wilson's disease; deficiency symptoms listed below.

Amounts

RDA: 15 mg men and women.
Protective: 15–50 mg.
Emergency: 40–50 mg.

Supplements

15–50 mg.

Synergists

Vitamin A; vitamin B_6; vitamins C and E.

Limiting Factors

Stress; alcohol; oral contraceptives; excess copper, lead or cadmium.

Deficiency Symptoms

Weak sense of taste; white spots on fingernails; diabetes or hypoglycemia; mental imbalance; slow growth in children.

Food Sources

Whole grains; green vegetables; nuts and seeds; sea vegetables.

Summary of Suggested Supplements

Supplement	Protective	Emergency
Vitamins		
A	10,000 IU	10,000-25,000 IU
B_6	10 mg	10-100 mg
B_{12}	2,000 meg	2,000 meg
Pantothenic acid	—	low-dose B-complex
C	500-3,000 mg	3,000-10,000 mg
E	50-400IU	200-600IU
Minerals		
Calcium	—	any supplement *except* bone meal
Chromium	100-300 meg	100-300 meg
Iodine	no supplement necessary	150 mg potassium iodide
Iron	—	10-18 mg
Magnesium	—	if calcium supplement is used, take magnesium equalling half that dose
Selenium	100-200 meg	200 meg
Zinc	15-50 mg	40-50 mg
Other		
Cysteine	500-1,000 mg (take with three times this amount of vitamin C and some B_6)	2,000 mg
Garlic tablets (Kyolic®)	1,200 mg	(1,200-1,800 mg)

8

Recipes

The following recipes will introduce you to the key foods of the Menu for the Nuclear Age. They are a sample of the culinary possibilities offered by wholesome natural ingredients and are presented with the hope that they will inspire you to discover and create new dishes on your own. Trying some of the cookbooks listed in the Appendix will be a good next step toward developing facility and pleasure in the use of these foods. And you will have the added satisfaction of knowing you are doing everything you can to enhance your health.

The majority of these recipes are designed to serve four people and the preparation time for each is given. The ingredients are kept simple so that you won't need to stock too many new foods. It's easy to eat right!

One thing that will ensure that you find this way of eating easy to adopt is to make sure you have a few essential items in your kitchen. If you don't have them already, purchase at least two heavy stainless steel or cast iron 2-quart pots, a medium-sized iron skillet, a good-sized soup pot, and a vegetable steamer basket. Of course, you can use pots and pans you already own, though it is recommended that you avoid aluminum cookware.

A couple of good sharp knives and a cutting board are important. So are some quart-sized (or larger) glass jars with lids or stoppers, for storing grains, beans and sea vegetables. You may wish to acquire these gradually.

I would like to add a note about cooking with oil. As we have seen, it is important to reduce fat intake. Nevertheless, some recipes require sautéing. In this case the thing to do is to use very little oil, just barely covering the bottom of the pan. On a day-to-day basis, steamed vegetables are preferable to sautéed or stir-fried. A sauce from the section on sauces and dressings may add something you feel is missing from vegetables that are not sautéed in butter or drenched in butter sauce. Salt is an important flavoring ingredient and it is good to add a pinch of salt per cup of cooking grain to all recipes. Good-quality natural sea salt, without any additives, is recommended. If you place a few grains of rice in the bottom of the salt shaker, the salt will be easy to pour even in damp weather.

Note: After Fukushima it is suggested to avoid seaweed and other foods from Japan

Whole Grains

Ideally, grains-that is, whole grains, not flour products—should make up 30 to 40% of your daily diet. For maximum benefit in the case of a radiation accident or spill of waste, increase the proportion of grains also increase the amount of miso, sea salt and sea vegetables in your diet, and avoid sugar *entirely.*

Cooking Grains

There are two good cooking methods for whole grains. Whichever method you choose, a heavy pot with a cover is a necessity. Also, be sure to wash the grain thoroughly and allow the excess water to drain before cooking. The first method for cooking whole grains (except millet) is to place the desired amount in the pot and add water. Bring to a gentle boil, then cover the pot. Reduce the flame and cook the appropriate amount of time. The second way is to heat a small amount of oil in the bottom of the pot. Add grain and saute for a few minutes before adding boiling water. Cook (as described above) for the appropriate amount of time. This method results in drier, more separated grains.

The instructions given here are general instructions. If a recipe gives different information, follow those particular instructions.

The proportions listed on the following page will yield two to three servings.

Add salt before cooking.

Suggested Water Requirements and Cooking Times for Grains

Grain	Water	Time
1 cup barley	2 ¼ cups	45 minutes
1 cup brown rice	2 cups	45-60 minutes
1 cup bulghur wheat	2 cups	20 minutes
1 cup kasha	2 ¼ cups	45 minutes
1 cup millet	2 cups	5 minutes; then remove from heat, cover, and allow to sit 20 minutes
1 cup rolled oats	2 cups	20-30 minutes (depending on variety)
1 cup scotch or steel-cut oats	3 cups	soak in cold water over-night; simmer 20-25 minutes, uncovered

Breakfast Oatmeal

There are three basic kinds of oats: quick-cooking or rolled oats (which have been heated during processing), steel-cut or scotch oats (which have not undergone the high temperatures required when oats are rolled), and whole oats. Steel-cut or whole oats are richest in B vitamins and are recommended for making oatmeal.

SERVES: 4 TIME: soak overnight; cook 20 minutes

1 cup steel-cut or scotch oats
3 cups water pinch of sea salt
¼ cup toasted chopped almonds or walnuts (optional)

Put oats and water in saucepan; cover and soak overnight. In the morning add salt, bring to boil, and simmer for 20 minutes, uncovered. Stir occasionally. Just before finishing cooking, add toasted nuts if desired. Instead of soaking oats overnight, you can cook them for 1 hour.

Croquettes

SERVES: 4   TIME: **under 40 minutes**

1 cup millet
½ cup sunflower seeds (hulled)
½ cup grated carrot
¼ cup minced scallions
¼ cup minced parsley
2 tablespoons soy sauce (or to taste)

Cook millet 5 minutes and then let sit off fire for 20 minutes. Bake sunflower seeds on cookie sheet in 350° F oven for 4 or 5 minutes and then let cool a few minutes. Crush with rolling pin or grind in seed or coffee bean grinder. Mix seeds, carrot, scallions and parsley with millet and stir in soy sauce to taste.

Shape mixture into patties and pan-fry in heavy skillet, lightly brushed with oil, for 5 minutes. These croquettes are very good with baked squash and steamed greens.

Brown Rice and Barley with Sunflower Seeds

This interesting recipe combines two grains. The addition of sunflower seeds balances the protein and adds zinc.

SERVES: 4   TIME: **50 minutes**

¾ cup brown rice
½ cup barley
2 cups water
¼ cup sunflower seeds

Cook rice and barley in heavy saucepan for 45 minutes. In the meantime, toast sunflower seeds in heavy pan for a few minutes, being careful not to burn them. Put toasted sunflower seeds in bowl and set aside. When rice is cooked, stir in seeds and serve.

Optional: Add 2 tablespoons of chopped parsley to the cooked rice. Cover saucepan and let it sit a minute to partly steam the parsley. Then stir together and add toasted sunflower seeds.

Brown Rice Salad

SERVES: 4 TIME: under 1 hour

2 celery stalks
1 carrot
2 scallions
½ cup chopped watercress
1 tablespoon chopped parsley
¼ cup chopped almonds or walnuts (or other nuts)
2 cups cooked brown rice

Chop celery stalks into bite-sized chunks, shred carrot with potato peeler, and chop scallions into one-inch pieces. Remove heavy stems from watercress and chop. Mix vegetables and chopped nuts in salad bowl and fold in cooked rice. Stir in Lemon-Miso Salad Dressing (see page 232) or another dressing of your choice just before serving.

Brown Rice with Wild Rice

SERVES: 4 TIME: 50 minutes

1 cup brown rice
½ cup wild rice
2 ½-3 cups water pinch of sea salt per cup of grain

Wash brown and wild rice and place in saucepan with water. Add sea salt. Cover saucepan and bring to boil. Simmer for 45 minutes, then remove from flame. Let rice sit for about 4-5 minutes before transferring it to serving bowl. Garnish with chopped parsley or scallions and serve.

Cooking tip: You may want to prepare a little extra of this dish, because if there is some left over you can reheat it with some vegetables for lunch the following day.

Fried Rice with Cabbage and Chick Peas

This combination of rice, beans and vegetables makes a delicious and easy meal in a dish. According to your taste and what you have available you could add some lightly fried tofu or tempeh, or some toasted and crushed almonds or walnuts. You might want to experiment with other combinations of vegetables and beans. Leftover beans lend themselves well to this dish.

SERVES: 4 TIME: under 1 hour

1 teaspoon sesame or corn oil
1 clove garlic, minced
½ cup toasted chopped almonds
1 cup finely chopped cabbage
½ cup cooked chick peas
1 ½ cups cooked brown rice
1 ½ teaspoon miso
2 tablespoons water

Brush heavy frying pan with oil. Add garlic and almonds and sauté for 2 or 3 minutes until browned. Add chopped cabbage and cook, covered, for 5 minutes. Add chick peas and rice and heat through, stirring gently. Blend miso with water in a cup, then stir into rice mixture. Remove from heat and let sit a few minutes to allow flavors to blend, and serve.

Lemoned Rice and Barley

SERVES: 4 TIME: 50 minutes

2 cups stock or water pinch of sea salt
1 clove garlic, crushed or chopped
¾ cup brown rice
¼ cup barley
1 tablespoon grated lemon rind
2 tablespoons fresh dill
1 tablespoon chopped fresh parsley

Heat soup stock or water, salt and garlic in heavy saucepan to boil. Stir in grains, cover, and simmer about 45 minutes. Remove from heat. Add grated lemon rind and let saucepan stand covered for a few minutes. Stir in dill and parsley and serve.

Millet with Crushed Almonds

SERVES: 4 TIME: 25 minutes for millet; about 50 minutes for rice

1 cup millet
2 cups stock pinch of salt
½ cup almonds (shelled)
1 teaspoon soy sauce

Cook millet. In the meantime, roast almonds on cookie sheet in 350° F oven for about 7 minutes. You will know that almonds are done enough when they begin to glisten. Place them in bowl and after they have cooled a few minutes sprinkle them with soy sauce. Let almonds dry, then crush with rolling pin until coarse. Put cooked millet in bowl and cover with a layer of crushed almonds. For variety, use 1 cup brown rice in place of millet in this recipe.

Rice Porridge with Vegetables and Plum

This porridge is a meal in a dish. Umeboshi, the salted plum, provides zesty seasoning and enhances the flavor.

SERVES: 4 TIME: under 1 hour

2 cups diced carrot, daikon radish and broccoli, or selection of your favorite vegetables
3 salted umeboshi plums, minced
2 cups stock or water
2 cups rice
¼ cup finely chopped parsley
¼ cup scallions, chopped in one-inch pieces

Cook rice for 45 minutes. When nearly done, boil vegetables and minced plums in stock. When they are just about tender (approximately 5 minutes), add rice, cover and cook over low heat for several minutes. Just before serving, stir in parsley and scallions. Cover pan and let sit off the fire for a minute.

Cooking tip: Vary the amount of liquid here according to your preference.

Pasta

Tradition says that Marco Polo introduced pasta to Italy after his journeys in China. The Chinese developed the long thin noodle centuries before and the Italians varied this "paste" of flour and water, making the hundreds of types and shapes of pasta that are popular today.

Pasta can be made of durum wheat, whole wheat plus bran, or wheat plus a variety of vegetables such as spinach. Pasta can also be made from other grains, such as rice or buckwheat. Buckwheat pasta is usually available in the form of *soba noodles.*

Although pasta can be a satisfying addition to the diet, it should never be substituted for whole grains on a day-to-day basis. Whole grains are the mainstay of the Diet for the Atomic Age.

Pasta Variations

Experiment with different kinds of pasta or noodles and try them with various sauces. Here are a few ideas:

- Olive oil heated with minced fresh basil.
- Sauteed pine nuts with garlic.
- Chopped black olives, cubed tomatoes and slivered toasted almonds or walnuts with sliced scallions.
- Olive oil with freshly ground black pepper and finely chopped Italian parsley.
- Olive oil heated with crushed rosemary, minced onion, minced garlic and lightly steamed green beans, sliced diagonally into one-inch pieces.

Cook and drain pasta and then toss with any of the above.

Spicy Sesame Noodles

This is one of my favorite dishes when I go to a Chinese restaurant. Be sure not to overcook the noodles. You can vary the spiciness by using more or less chili powder. The scallions add color and crunch.

SERVES: 4 TIME: 20 minutes

4 quarts water
1 teaspoon sea salt
1 pound buckwheat noodles
2 teaspoons white miso
¾ cup tahini, blended with ¼ cup water

2 teaspoons Szechwan hot chili oil or 1 teaspoon chili powder
8 scallions, cut diagonally into 1-inch pieces

Bring 4 quarts of water to boil in a big pot. Add salt, drop in noodles and cook for 6 to 12 minutes. Drain cooked noodles in colander and rinse under cold water.

Mix miso with 2 tablespoons water. Then mix with tahini and chili. Transfer drained noodles to large bowl, add scallions and pour in tahini mixture. Toss well. Either serve warm or chill and serve later.

Muffins

Cornmeal Muffins

There's nothing like delicious corn muffins. They are a real treat hot out of the oven for breakfast—or on hand to go with a simple lunch or dinner.

MAKES: 12 MUFFINS TIME: 30 minutes

1 cup cornmeal
1 cup whole wheat flour
½ teaspoon salt
2 teaspoons baking powder
1 tablespoon honey
2 tablespoons vegetable oil
1 egg
1 2/3 cups tepid water

Preheat oven to 375° F. Combine cornmeal, whole wheat flour, sea salt and baking powder. Add honey and oil to water; blend liquid mixture and then stir into flour. Beat egg slightly and add to mixture. Grease 12-cup muffin tin and fill each cup three-quarters full. Bake at 375° for 25 to 30 minutes, until muffins are toasty looking on top. Let cool a bit before eating.

Vegetables

What is more sweet and delicious than vegetables out of your own garden? If a home garden isn't practical for you, try to find a farmers' market nearby. The important thing is to find a convenient source for the freshest vegetables possible. The produce that they call "organic" is excellent, but is not a "must." Just be sure to wash vegetables very well.

Vegetables are an important part of the Menu for the Nuclear Age because of their high vitamin and mineral content and detoxifying properties. They should comprise about one-quarter of your food intake each day. A steamer that fits inside of a pot is necessary kitchen equipment—steamed vegetables are simple to prepare, tasty and most nutritious. Add a sauce of miso (with or without some tahini, lemon or other flavoring), and a side dish of brown rice or tofu, and you have a fine meal. Try to incorporate six or seven different vegetables into your cooking each day.

Carrots and Onions

SERVES: 4 TIME: 10 minutes

1-2 teaspoons sesame oil
1 tablespoon pressed garlic
1 tablespoon fresh grated ginger
1 or 2 onions, chopped
2 carrots, sliced into matchsticks
1 tablespoon ground sesame salt

Cover bottom of heavy pan with oil and heat. Add garlic and ginger and cook for a minute or so. Add onions and carrots and sauté 2 minutes. Add 2 tablespoons water, cover and steam 5 minutes. Sprinkle with sesame salt.

"Corn-off-the-Cob" with Umeboshi

Easier to serve and handle than corn-on-the-cob, "corn-off-the-cob" also does away with the need for butter. Umeboshi plums add such terrific flavor and zip[1].

SERVES: 4 TIME: 8 minutes

5 or 6 ears of fresh sweet corn
4 umeboshi plums, minced water

Bring water to boil. Remove husks and wash corn quickly with cold water. Place in boiling water, cover, and reduce flame to medium-low. Simmer 3-4 minutes. Remove corn and cut it off cob with sharp knife. Place in serving bowl and mix in minced umeboshi plums.

Green Beans with Lemon Miso

Lemon miso is also delicious with zucchini or brussels sprouts.

SERVES: 4 TIME: 10 minutes

1 pound green beans
2 teaspoons miso
2 tablespoons hot water
2 tablespoons lemon juice
2 tablespoons finely chopped parsley

Trim and slice green beans and steam in steamer for about 5 minutes. Blend miso with hot water; add lemon juice and parsley. In a serving bowl, toss green beans with miso mixture. You can add sunflower seeds if you like.

Mustard Greens or Kale with Red Onions

SERVES: 4 TIME: 10-15 minutes

1-2 teaspoons sesame oil
2 red onions, chopped
4-6 cups mustard greens or kale, sliced diagonally
3 tablespoons water
2-3 tablespoons sesame seeds

Brush bottom of heavy pan with oil. Allow oil to heat, then saute onions a few minutes (until translucent). Add greens and cook a few minutes. Add 3 tablespoons water, cover and cook several minutes longer. Vegetables should remain slightly crunchy. Sprinkle with sesame seeds.

Old-Fashioned Vegetable Stew

Serve this delectable dish over some steaming brown rice or millet, with Corn Muffins (see page 202), or with sourdough bread.

SERVES: 6 TIME: 30 minutes

4 leeks, sliced in ½-inch pieces
4 carrots, sliced diagonally
½ head red cabbage
2 cups vegetable stock
1 pound tofu
1 tablespoon grated ginger
2 or 3 scallions

Brush oil in heavy pot and heat. Add leeks, carrots and cabbage. Add 2 cups stock, cover, and cook 20 minutes. In the meantime, cut tofu into cubes and fry in oil and ginger till well coated. When vegetables are cooked, stir in tofu. Serve garnished with chopped scallions. This stew is excellent when reheated.

Oriental Cabbage and Celery

SERVES: 4 TIME: 15 minutes

1 teaspoon sesame oil
½ head cabbage, chopped
2 or 3 stalks celery, chopped
1 tablespoon umeboshi plum paste
2 teaspoons miso
2 tablespoons water

Brush oil in heavy pan and heat. Add cabbage. Cook for a minute or two, then add celery. Stir, cover, and cook about 10 minutes. Add umeboshi paste and cook about 2 minutes more over low flame. In a small cup, blend miso with water. Stir liquefied miso in with vegetables. If you wish, you can omit miso and instead add 1 tablespoon plum paste after vegetables are cooked.

Picnic-Style Cole Slaw

Whether or not you take this cole slaw to a picnic it is still satisfying and fortifying.

SERVES: 4 OR 5 TIME: 10 minutes

3 cups finely chopped cabbage
1 ½ cups grated carrot
½ cup sunflower seeds (hulled)
½ cup raisins
½ cup alfalfa sprouts
1 cup tofu

2 teaspoons miso
2 tablespoons water
3 tablespoons lemon juice
½ teaspoon mustard
1 tablespoon vegetable oil

Steam cabbage and carrots lightly. Drain steamed vegetables; mix with sunflower seeds, raisins and alfalfa sprouts in big bowl. Beat tofu either by hand or in blender until thick and creamy. In small bowl or cup, add some water to miso to liquefy it. Add liquefied miso, lemon juice, mustard and oil to tofu. Mix this dressing well with the salad before serving.

Quickly Boiled Watercress and Carrots

This is a very quick vegetable dish that combines two—powerful vegetables and a brilliant color combination.

SERVES: 4 TIME: 5 minutes

1 carrot, thinly sliced on a diagonal
2 bunches of watercress

Place 1 inch of water in pot and bring to boil. Cook carrots for 50-60 seconds. Use slotted spoon to remove carrots, leaving water in pot. Drain carrots in colander. Simmer watercress for 50 seconds, stirring or mixing quickly to cook evenly. Remove watercress, drain, and place on cutting board. Slice into 2-inch pieces. Mix carrots and watercress in serving bowl; serve with your favorite dressing.

Sauteed Broccoli

SERVES: 4 TIME: 10 minutes fresh broccoli
1 teaspoon sesame oil

1 tablespoon fresh grated ginger
1 tablespoon pressed garlic
1 tablespoon sesame salt

Wash broccoli well. Trim off leaves and save them for soup stock. Peel tough stems and chop in small pieces to obtain about 3 cups of flowerets and chopped stems. Heat oil in heavy pan. Add ginger and garlic. After a minute or two, add broccoli and sauté about 5 minutes, stirring occasionally. Cover and steam a few minutes, until stems are just tender. Sprinkle with sesame salt just before serving.

Scrumptious Squash Puree

Taste, simplicity, brilliant color and high vitamin A content recommend this easy vegetable dish.

SERVES: 4 TIME: 1 hour and 10 minutes

1 butternut squash, about 3 pounds
½ teaspoon ground cardamom
¼ teaspoon ground cloves
½ teaspoon sea salt
1 teaspoon umeboshi plum paste

Cut squash in half and cook in 375° F oven for about an hour or until tender. Remove from oven and let cool. Then spoon squash out of skin; puree in blender or food mill or mash with potato masher. Combine seasonings and mix them in. Gently heat mixture in heavy saucepan and stir until heated through.

Steamed Broccoli

Steamed broccoli is delicious "just as is" or with sauce. It is best when cooked tender but still a bit crisp.

241

SERVES: 4 TIME: 15 minutes

4 cups broccoli flowerets water

Place about 1 inch of water in pot. Set steamer down inside pot. Place broccoli in steamer, cover pot, and bring to boil. Reduce flame to medium-low and steam broccoli several minutes, until it is tender but still firm and bright green. Remove and place in serving bowl.

Steamed Greens

The principle for steamed greens is the same as for steamed broccoli.

SERVES: 4 TIME: 5-10 minutes

1 bunch collard greens, kale, or other dark leafy greens

Put about an inch of water in pot and place steamer inside pot. Bring water to boil. Rinse greens very well. Slice them lengthwise once or twice and then slice crosswise into thin pieces. Place greens in steamer, reduce flame to medium-low and cover pot. Steam for about 3 minutes, until tender but still crisp and bright green. Remove and place in serving dish. For added flavor, you can toss greens with some lemon juice or soy sauce. Umeboshi paste is also delicious with greens. It adds a distinctive, engaging flavor.

Summer Salad

Try this summer salad with quick-steamed watercress and rye bread. You can also vary the vegetables or add a little leftover rice.

SERVES: 4 TIME: 10 minutes

1 cup lightly steamed green beans
1 cup lightly steamed carrots, cut in matchsticks
¼ cup finely chopped parsley
¼ cup tahini
3 tablespoons lemon juice
1 ½ teaspoons miso
2 tablespoons water
2 cloves garlic, pressed

Place vegetables in large salad bowl. Blend tahini, lemon juice, miso, water and garlic for dressing. Add dressing and toss salad just before serving.

Beans

A concentrated source of B vitamins, minerals and protein, beans ideally make up 5 to 10 percent of the diet. It is good to have one serving of beans daily. They complement the protein in grains. Chick peas should be soaked overnight before cooking. Aduki beans can be soaked about four to six hours. Always rinse beans well and remove any discolored or odd pieces before soaking. Just before cooking, place a short (1 inch or less) strip of kombu in the bottom of the pot. This adds flavor and makes the beans more digestible. As a general rule, one cup of beans will yield about two cups of cooked beans.

Times for Cooking Beans

Beans	Time
1 cup chick peas	1 ½ hours
1 cup lentils	¾ hour
1 cup aduki beans	1 hour
1 cup soybeans	2 hours
1 cup split peas	¾ hour

Aduki Beans with Squash

This dish should be cooked in a heavy saucepan and not stirred while it is cooking. In this way the flavors will meld and the sweetness of the squash will pervade the dish.

SERVES: 4 TIME: 1 hour

1 strip kombu, about 6 inches long
1 cup aduki beans, soaked 8 hours
5 cups water or stock
1 cup cubed butternut squash
¼ teaspoon sea salt
1 teaspoon soy sauce

Place kombu in bottom of heavy saucepan and cover with beans and water. Bring to boil, then cover saucepan, lower heat and simmer 30 minutes. Layer cubed squash on top of beans, add sea salt and cook 20 minutes more. Add soy sauce and stir gently to blend.

Bean Curry

This dish is tasty with Lemoned Rice and Barley (see page 199) and a steamed green vegetable.

SERVES: 4 TIME: soak overnight and cook 45 minutes

1 piece of kombu, 1 inch long
1 cups kidney beans
8 cups water or stock cloves garlic, minced
1 onion, chopped
1 teaspoon sesame or corn oil
½ teaspoon sea salt
¾ teaspoon curry powder (or to taste)

Rinse and clean beans with water. Soak 8 hours, then discard water used for soaking. Place kombu in bottom of heavy saucepan and add beans. Cover them with water or stock and bring to boil. Lower heat; simmer for 45 minutes. Drain. In the meantime, sauté onion and garlic with oil in saucepan for several minutes, until translucent. Add drained beans, sea salt and curry powder and cook for a few minutes to blend flavors.

Chilled Chick Pea Salad

If you have any leftovers when you make this hearty salad, they will be welcome tomorrow!

SERVES: 3 OR 4 TIME: soak overnight and cook
for approximately 1 ½ hours

¾ cup chick peas
3 tablespoons finely chopped parsley
3 tablespoons onion, chopped in small pieces
2 or 3 scallions, chopped
1 teaspoon miso juice of 1 lemon

Soak chick peas overnight. Drain, cook, and allow to cool. Then add vegetables and sauce made of miso and lemon juice. Chill salad and serve as a side dish.

Lentil Salad

Lentils do not require presoaking, so this is an ideal quick bean dish. Flavorful and fortifying, this salad will extend your culinary horizons if you've only tasted lentils in lentil soup. Lentil salad makes a terrific luncheon dish at any time of year.

SERVES: 3 OR 4 TIME: about 45 minutes

¾ cup lentils
3tablespoons finely chopped red onion
2 tablespoons finely chopped parsley
4 garlic cloves, pressed
2 tablespoons umeboshi vinegar
2 teaspoons white miso
2 tablespoons water

Be careful not to overcook lentils; they should remain firm. Drain lentils and put them in bowl. Add onion, parsley, garlic and vinegar. In small cup, blend miso with water. Stir mixture into salad. Serve at room temperature.

Vegetarian Chili

When I first sampled this dish I found it surprisingly delicious. Bulghur replaces the usual ground meat. It is not missed. Try this recipe and you will see.

SERVES: 4 TIME: 35 minutes

2 tablespoons olive oil
1 tablespoon minced garlic
2 ½ cups finely chopped onion
1 ½ cups chopped carrots
1 tablespoon ground cumin
1 teaspoon chopped fresh basil
1 teaspoon dried oregano
1 teaspoon sea salt
1 ½ cups vegetable stock
¾ cup bulghur
1 cup chopped tomatoes
2 cups cooked kidney beans
½ teaspoon Tabasco (hot pepper sauce)
2 tablespoons lemon juice
2 tablespoons chopped green chilis

Heat olive oil in large frying pan. Add garlic and sauté briefly. Then add onions and sauté a few minutes more. Add carrots, celery, spices and stock. Stir to blend. Add bulghur and tomatoes and cook 20 minutes. Then add kidney beans, Tabasco, lemon juice and chilis. Stir and simmer a few more minutes.

Afterword

The world we live in will not get better unless we do some things to make it better. The fact that you have just read this book shows that you are concerned enough to learn more about the growing problem of radiation and its possible solution. The next step is to take what you have learned and use it to improve the quality of your life and the lives of those you care about.

If we all do this, maybe we can then tip the scales back in favor of humanity.

For more impetus to tip the scales read: www.nuclearreader. info.

APPENDIX
2012

Low Level Radiation Hazards

This book focuses on the health damage of low level radiation and how to utilize foods as a radiation shield to the extent that this is possible within the context of the increasing global levels of radioactivity. The following is a discussion of some of the general effects of low level radiation aside from specific health damage.

Low level radioactivity includes radiation released from the routine operation of the world's 433 nuclear power plants, plus leaks and accidents, nuclear fuel recycling, uranium mining and other related activities.

FIVE POINTS OF CONSIDERATION

INFERTILITY

Radiation causes infertility. The global fertility rate has dropped by nearly half since 1955.

WEAKENED
IMMUNE
SYSTEM

Radiation weakens the immune system. A hundred nation study on quality of health found the United States was number 1 in 1943. By 1992, the United States was number 100, according to a U.S. Public

Health Statistics Report. Globally, health is deteriorating with the incidence of cancer, heart disease, allergies and infectious diseases increasing.

MUTATED VIRUS AND BACTERIA

Even at low levels radiation may increase mutations of bacteria and virus. Mutations are causing the appearance of new diseases such as Reye's Syndrome, Legionnaires' Disease and Lyme Disease.

LOSS OF OXYGEN GLOBALLY

The percentage of oxygen in the air is down to about 19 percent. The standard reference amount is 21 percent. Oxygen is formed by trees and plankton. Trees and plankton can be killed by radiation.

OZONE BREAKDOWN

Large-scale breakdown of the protective ozone layer in the stratosphere was initiated in 1958 by high atmosphere bomb tests, and continues due to releases from nuclear power plants and reprocessing plants. Radioactive Krypton-85 goes to the stratosphere where it greatly enhances CFC ozone damage.

INFERTILITY

The global fertility rate has dropped by nearly half since 1955! The cumulative effects of radiation-caused infertility raise the possibility of gradual human extinction.

Escalating infertility in the United States has forced couples to turn to the fast emerging new world of assisted reproduction and pre-made embryos in ever growing numbers. A front page *New York Times* article (November 23, 1997, "Clinics Selling Embryos Made for Adoption") looks at the "anguished infertile couples" who "are more than willing to pay for whatever infertility clinics can offer."

Fear of the "population bomb" of the 1960's has turned into the "birth dearth" of the 1990's.

The so-called replacement rate is 2.1 children, which is needed to keep the population from falling. In the year 2010 fertility rates hovered between 1.1 and 1.4. In the less developed countries fertility rates were 6.0 in 1972 and had dropped to 2.9 in 2010. The current fertility rate in the developed nations is 1.6 children per woman. According to the CIA World Factbook, 2010, the fertility rate of the United States has nearly halved in the last 50 years.

Dr. John Gofman, an eminent scientist, medical doctor and eloquent spokesman against the hazards of nuclear power explained, back in the 1970's, "that the worry about over-population would become a non-worry due to radioactivity." (Tamplin, Arthur R. and John Gofman. *Population Control Through Nuclear Pollution.* Chicago: Nelson-Hall, 1970.)

This prediction is confirmed in the decline in fertility among those born during the period of atmospheric bomb testing between1955 and 1963.

Dr. Rosalie Bertell, mathematician, epidemiologist and founder of the International Institute of Concern for Public Health has been researching infertility for some years and feels it is the "cutting edge" of radiation health damage, surpassing immune damage in the extent of its implications, as it raises the possibility of human extinction.

In response to questions on the status of this research, Dr. Rosalie Bertell wrote in a personal correspondence (November 1997): "We found in Kerala, India, that when comparing couples matched for socioeconomic status, class, religion, occupation and life style, those living on the high radiation background (300 to 3000 mrad per year) had twice the infertility rate of those living on the normal background soil (below 300 mrad per year)."

The Baby Boomers are the group born in the USA after the war, from 1945 through 1963, and these are the years of atmospheric bomb testing as well as the start of the nuclear power industry. They show a high rate of immune related diseases and also an increasing rate of infertility. Data from the U.S. Public Health Service illustrate the difference between the fertility rates of the Baby Boomers and those who are called Pre-Baby Boomers:

Percent Women Infertile, by Age
U.S. 1965 and 1976

	Percent Infertile in 1965	Percent Infertile in 1976	Percent Change
Baby Boomers			
Age 15-19	0.6	2.0	+1.4
Age 20-24	3.4	5.6	+2.2
Age 25-29	6.1	8.4	+2.3
Pre-Baby Boomers			
Age 30-34	10.8	9.5	-1.3
Age 35-39	13.4	11.4	-2.0
Age 40-44	18.5	14.6	-3.9

(Reproductive Impairments Among Married Couples United States U.S. Public Health Service, Washington, D.C., December 1982)

The two surveys in the chart, done in 1965 and 1976 by the US Public Health Service, show that the percent of infertility of the baby boomers increased, and the percent of infertility of the pre-baby boomers born before bomb testing and nuclear power, decreased.

In an article in the conservative journal *Foreign Affairs* Nicholas Eberstadt says:

"If the twentieth century was marked by vast improvements in public health, then the twenty-first century is likely to be defined by steep declines in fertility rates." He points out this new demographic reality includes the West, Europe and rising economies. (Eberstadt,

Nicholas. "The Demographic Future: What Population Growth Means for the Global Economy." *Foreign Affairs 89, no. 6* Nov/Dec 2010).

WEAKENED IMMUNITY

On the correlation between low dose radiation and weakened immunity radiation physicist Dr. Ernest Sternglass stated in a 1986 article: "It appears that perhaps the most serious unanticipated effects of fallout is long-term, persistent immune deficiency." He adds, "It can weaken the immune defenses of the body at very low total doses leading to unexpectedly large increases in infectious diseases and cancers." (*Int. J. Biosocial Res.,* July 1986, p. 18)

Initially, zealous industry misrepresentation of the facts led the public to believe that small amounts of radiation were of no special concern. Yet these low levels are exactly the cause of weak immunity and resulting diseases.

Authorities couldn't ignore emerging data, and in December 1989 the government sponsored National Academy of Sciences, in a report titled 'Biological Effects of Radiation', stated that there was no safe level of radiation.

Low protracted doses of radiation cause physiological damage through the formation of free radicals. A free radical is a molecule with an imbalance in electrons which can destabilize other molecules resulting in cellular damage and disease.

In high, short doses like the Hiroshima bomb blast, radiation primarily causes direct damage to the nucleus of cells, where the genes that control the functioning of the cell are located. In contrast, low doses acting continuously over time produce their damage indirectly through the generation of free radicals that destroy cell-membranes; hundreds to thousands of times more efficiently than might be expected in calculations related to high-dose damage. So the everyday amount of radiation that is released as part of the normal operation of the world's 400 nuclear power plants is a very grave concern. Nuclear power plants release steam and water as part of their normal operations, and these releases, even though they may be partially filtered, disperse radiation into our air and drinking water, onto farmland and into our food.

The everyday releases of low-level radioactivity by nuclear power plants has been found to cause several kinds of health damage; including premature births, congenital defects, infant mortality, mental retardation, heart ailments, arthritis, diabetes, allergies, asthma, cancer, genetic damage and chronic fatigue syndrome. It has been linked to previously unknown infectious diseases and the resurgence of old ones by damaging the developing white blood cells originating in the bone marrow and thus weakening the immune system.

For more information on health damage see chapters 3, 4 and 5.

MUTATED BACTERIA AND VIRUS

It is well known that radiation can cause mutations in bacteria and viruses. Andrei Sakharov, the famous Russian physicist, described in his 1992 Memoirs that even at low levels radiation could increase mutations of bacteria and viruses. His predictions, which were originally made in 1958, have come true and we are seeing new ailments such as Reye's syndrome which first appeared in 1963, and Legionnaires' disease, which is caused by a bacteria that was not threatening prior to 1976. AIDS may be related to a mutated virus combined with a weakened immunity in a generation born after the first nuclear weapons were detonated.

Of particular interest is Lyme Disease which first appeared in 1975 near the Millstone and Haddam Neck nuclear power plants in Connecticut. Dr. Jay Gould points out:

> "In 1975 there were 59 cases of Lyme Disease recorded; in 1985 the number increased to 863, mainly in the two counties of Middlesex and New London, CT near the Millstone Nuclear Power Plant. Just as increases in cancer

may be linked to the huge radiation release from Millstone in 1975, so too may be the tick-borne Lyme disease epidemic. The Lyme Disease is carried by a spirochete that had not been as harmful to humans prior to 1975. It is well known that radiation can cause mutations in bacteria. The enormous 1975 Millstone radiation release may have caused just such a mutation in the tick-borne spirochete." (Gould, J and B Goldman. *Deadly Deceit: Low-Level Radiation and High-Level Cover-Up.* New York: Four Walls Eight Windows, 1990.)

So we have a double challenge: the weaker immune system and the new diseases resulting from mutated pathogens.

Dr. Ernest Sternglass further explains:

"When the radiation from such isotopes as strontium-89 and 90 in the bone marrow mutates an existing virus that invades the T-cells of the immune system and kills them in the process of replication, the stage is set for the complete collapse of the immune defenses and resulting death from opportunistic infections or cancer." (Ernest Sternglass, "The Implications of Chernobyl for Human Health", *International Journal of Biosocial Research*, no.8, p. 19, July 1986)

LOSS OF OXYGEN GLOBALLY

Walter Russell, a visionary artist and scientist predicted in his book *Atomic Suicide?*, published in 1957, that due to man-made radioactivity we would suffer a reduction of oxygen in the atmosphere. Like the predictions of Andrei Sakharov in the 1950's, Walter Russell's foresight is now coming true. Our current oxygen levels are down to about 19 percent (*BioTech News* 1997). The standard reference amount is 21 percent oxygen. Some experts say that we may have originally evolved in an atmosphere of 38 percent oxygen. Now, due to the loss of the two main producers of oxygen, forests and ocean plankton, measurements as low as 12 percent and 15 percent have been made in heavily industrialized areas. This oxygen-depleted condition is a contributing cause of the generalized lack of well-being that many are experiencing. We need oxygen to live and to thrive!

Trees and other land plants provide about half our oxygen, and plankton provides the rest. Phytoplankton, the base of the marine food chain, is declining. Various studies confirm this: Plankton in parts of the Antarctic Ocean had already declined by up to 12 percent when S. Weiler testified to the Senate Commerce Committee in November of 1991. Trees absorb radioactive carbon-14 in place of stable forms of carbon and in this way they are gradually killed. In *The Petkau Effect*, Ralph Graeub describes how radioactivity has harmed trees and forests:

> "It is assumed that the decisive physiological damage resulting in current forest death must have begun during the 1950's. This is depicted in a reduction in density and width of tree rings, and in reduced growth, which is true in the Northern Hemisphere and in the Himalayas . . . . Neither aging, location, nor climate can be considered as the possible sole cause of damage . . . . The growth ring of a tree shows exactly what effects the tree has experienced, both in terms of time and seriousness . . . . During the 1950's and 1960's, there must have been a global wave of air pollution which caused the initial damage."

The author speculates that it may not be just the usual chemicals which are so damaging to trees. And he notes that these trees are mainly within the 30th to 60th parallels of northern latitude. "This zone contains the most nuclear power plants—over 300—and almost all nuclear reprocessing centers. Also, the vast majority of nuclear weapons tests occurred in this area."

OZONE BREAKDOWN

The protective layer of ozone around the Earth filters out solar and cosmic rays and prevents them from reaching our planet. Ozone surrounds the Earth in a layer between six and thirty miles above sea level. It is formed when light rays strike molecules of oxygen, which is O_2, and causes them to break into two separate oxygen atoms, or an O and O. An atom of oxygen then combines with a molecule of oxygen and forms ozone which is O_3. It breaks down again and then recombines again. And so on; unless it is interfered with. Radiation interrupts the process of ozone formation.

1957—Walter Russell published his book *Atomic Suicide?* whose principle message was that the development of the nuclear weaponry and nuclear industry, if it continued, would eventually destroy the planet's oxygen.

> "The element of surprise which could delay the discovery of the great danger, and thus allow more plutonium piles to come into existence, is the fact that scientists are looking near the ground for fallout dangers. The greatest radio-active dangers are accumulating from eight to twelve miles up in the strato- sphere. The upper atmosphere is already charged with death-dealing radio- activity, for which it has not yet sent us the bill. It is slowly coming and we will have to pay for it in another century, even if atomic energy plants ceased today."
>
> (Russell, Walter and Lao. *Atomic Suicide?*
> University of Science and Philosophy. Virginia 1957 p. 18)

1982 and 1984—Two German reports conclude that radioactive krypton, which is released in the daily operation of nuclear plants and through the reprocessing of used reactor fuel elements, is affecting the distribution of the electric fields in the atmosphere.

1987—The ozone hole is twice as large as the U.S. It is discovered that ozone is not only diminishing over the south pole but globally.

1987-1988—Consensus has it that various man-made chemicals are the sole cause of ozone breakdown; especially compounds of chlorine (CFC's) and bromine (from halon fire extinguishers) and there was an attempt to implicate hair spray and refrigerators. A leading authority on the ozone problem, NASA's Dr. Robert

Watson, admitted many scientists were "baffled" by findings of ozone depletion even in areas where CFC's action was negligible. He called the extent of the hole's growth "absolutely unexpected".

April 6, 1989—"Scientists reported yesterday that for the first time they have detected an increase in "biologically relevant" levels of ultraviolet radiation reaching the ground as a result of the ozone hole over the Antarctica." This is the first indication that the depletion of ozone is beginning to cause the potentially harmful effect that has long been predicted." *(The Washington Post 4/6/89)*

Late 1990—University of California researchers publish their findings that phytoplankton are reproducing less profusely than before. Observing the plankton in the Belingshausen Sea (in the Antarctic) they found that increased UV appears to be suppressing the phytoplankton's productivity by 6 to 12%.

1992—Both NASA and The World Meteorological Society reported 10 to 25% ozone depletion measured over the northern United States, Canada, Europe and the Antarctic; and the ozone hole is now three times the size of the United States.

1994—An article in a German journal *Strahlentelex* (March 3, 1994) argues that the nuclear industry is responsible for the hole in the ozone. The authors, Giebel and Sternglass explain that radioactive gases like krypton-85 from nuclear plants and from the recycling of spent fuel go up to the stratosphere where they create water droplets from the moisture which in turn form ice crystals which enhance the destruction of the ozone by the fluorohydrocarbons.

(Krypton-85 has a half-life of 10.7 years and a whole life of 217 years.)

March 1996—The World Meteorological Agency reports "the extremely worrying" development of an unprecedented 45 percent ozone thinning over Greenland, Scandinavia and Western Siberia.

Summer 1997—Research from the Antarctic Marine Living Resources Program find "krill abundance in the Antarctic Peninsula region is down 60 to 90 percent since the early 1980's".

●　　　●　　　●

Not much time remains before the chance to avert the threats we now confront will be lost and the prospects for humanity are immeasurably diminished. We, the undersigned senior members of the world's scientific community, hereby warn all of humanity of what lies ahead. A great change in the stewardship of the earth is required.

Warning to Humanity issued by the
Union of Concerned Scientists
signed by 1500 scientists
from 69 nations

In our every deliberation, we must consider the impact of our decisions on the next seven generations.

The Great Law of the
Iroquois Confederacy

Given all the damage to the Earth and to all living creatures caused by the nuclear industry, the natural response would be to find it unacceptable and to phase it out in favor of clean renewable energies. However, our attention has been diverted. The wrong questions are proposed. We respond to these questions and so miss the larger point.

We do not need to ask how we can meet the inflated western energy demands, but we should inquire how we can live appropriately without causing harm. We can stop producing more radioactive poison; phase out nuclear power plants and develop clean, renewable energy. We can clean up and store radioactive waste responsibly. If this is our true intention, it could be done. If not, we are destined for a radioactive Earth and all that entails.

In response to the March 2011 Japanese nuclear catastrophe, President Angela Merkel of Germany took seven power plants off-line and proposed the following six-point plan (*Spiegel Online International* 4/15/2011) from which other countries might take inspiration:

- **Expanding renewable energy.** Investing in more wind, solar, and biomass energies will try to raise the renewable-energy share of Germany's total energy use—from a baseline of 17 percent in 2010.

- **Expanding grids and storage.** Building a much larger storage and delivery network for electricity—particularly wind energy, which can be generated in the north but must be carried to the south—will be a main focus.

- **Efficiency.** The government hopes to improve the heating efficiency of German buildings—and reduce consumption—by 20 percent over the next decade.

- **"Flexible power."** The government wants to build more "flexible" power plants that can pick up slack from wind or solar energy when the weather fails to generate enough electricity during peak demand. The obvious source of "flexible power" for now, besides nuclear energy, is natural gas.

- **Research and development.** The government will increase government support for research into better energy storage and more efficient grids to a total of €500 million between now and 2020.

- **Citizen involvement.** The government wants to involve its sometimes-recalcitrant citizenry who are resistant to wind generators and the installation of an efficient new power line grid in some regions.

Websites:

A list of nuclear plants in the USA with their locations and details such as the amount of high-level radioactive waste on-site, problems and leaks, and "worst case" projections of casualties and property damage:
www.animatedsoftware.com/environm/no_nukes/nukelist1.htm

A list of 200 nuclear related books and videos:
www.animatedsoftware.com/environm/no_nukes/mybooks.htm

A national information center about nuclear power and sustainable energy issues:
www.nirs.org

A map of the USA depicting radiation levels updated frequently:
www.radiationnetwork.com

About the catastrophe in Japan:
www.consciousbeingalliance.com/2011/03/japans-catastrophe-nuclear-power-cover-up

Concise and comprehensive must reading:
www.nuclearreader.info.

The latest research:
www.radiation.org

About health effects of low level radiation:
www.nuclearreader.info
www.llrc.org   www.nirs.org
www.beyondnuclear.org   www.antenna.nl/wise
www.llcph.org

References

Preface

Balabukha, Vera, Ed. *Chemical Protection Against Ionizing Radiation.* Elmsford, NY: Pergamon Press, 1963.

Ball, Howard. Downwind from the Bomb. *The New York Times Magazine,* pp. 32-42, Feb. 9, 1986.

Caldicott, Helen. *Nuclear Madness: What You Can Do!* New York: Bantam Books, 1979.

Clark, Linda A. *Are You Radioactive?* New York: Pyramid Books, 1973.

Cummings, Judith. Three Decades After Bomb Test, Utah Sheep Ranchers Feel Little Bitterness. *The New York Times,* p. 18, Aug. 15, 1982.

Jacobs, Leonard. Natural Ways to Survive a Meltdown. *East West Journal,* p. 62, June 1979.

Schell, Jonathan. *The Fate of the Earth.* New York: Alfred A. Knopf, 1982.

Sternglass, Ernest J. *Secret Fallout: Low-Level Radiation from Hiroshima to Three Mile Island.* New York: McGraw-Hill, 1981.

Wasserman, Harvey and Solomon, Norma. *Killing Our Own: The Disaster of America's Experience with Atomic Radiation.* New York: Dell Publishing, 1982.

Chapter One

Abbots, John. Radioactive Waste: A Technical Solution? *The Bulletin of Atomic Scientists,* pp. 12-18, Oct. 1979.

Advisory Committee on the Biological Effects of Ionizing Radiation (BEIR), 1972 Report. *The Effects on Populations of Exposure to Low Levels of Ionizing Radiation.* Washington: National Academy Press, 1972.

Advisory Committee on the Biological Effects of Ionizing Radiation (BEIR), 1978 Report. *The Effects on Populations of Exposure to Low Levels of Ionizing Radiation.* Washington: National Academy Press, 1978.

Advisory Committee on the Biological Effects of Ionizing Radiation (BEIR), 1980 Report. *The Effects on Populations of Exposure to Low Levels of Ionizing Radiation.* Washington: National Academy Press, 1980.

Aoyama, Isao, et al. Evaluation of the Radioactive Wastes Disposal into the Deep Ocean. *Health Physics* 33:227-240, 1977.

Arkin, William M, and Fieldhouse, Richard W. *Nuclear Battlefields: Global Links in the Arms Race.* Cambridge, MA: Ballinger Publishing Company, 1985.

Ball, Howard. Downwind from the Bomb. *The New York Times Magazine,* pp. 32-42, Feb. 9, 1986.

Ball, Howard. *Justice Downwind.* New York: Oxford University Press, 1986.

Bartlett, Donald L, and Steele, James B. *Forevermore: Nuclear Waste in America.* New York: WW Norton & Co., 1985.

Berger, John J. *Nuclear Power: The Unviable Option.* Palo Alto, CA: Ramparts Press, 1976.

Bertell, Rosalie. X-Ray Exposure and Premature Aging, *journal of Surgical Oncology,* 9:379-391, 1977.

Bibliography of Reported Biological Phenomena on Chemical Manifestations Attributed to Microwave and Radio-Frequency Radiation. GIAZER, ZR Naval Research Institute Research Report, Project MF 12-524-015-004B, Report No. 2, 1971.

Broad, William J. Hans Bethe Confronts the Legacy of His Bomb. *The New York Times,* pp. C1-C2, June 12, 1984.

Brodeur, Paul. *The Zapping of America: Microwaves, Their Deadly Risk, and the Cover-Up.* New York: WW Norton & Co., 1977.

Bross, Irwin D. Hazards to Persons Exposed to Ionizing Radiation (and to their children) from Dosage Currently Permitted by the Nuclear Regulatory Commission, pp. 913-950 in *Effect of Radiation on Human Health: Health Effects of Ionizing Radiation,* vol. 1, Hearings before the Subcommittee on Health and the Environment of the Committee on

Interstate and Foreign Commerce of the U.S. House of Representatives. Serial No. 95-179. Washington: Jan./Feb. 1978.

Bross, Irwin D. Testimony presented in *Radiation Standards and Public Health—Proceedings of a Second Congressional Seminar on Low-Level Ionizing Radiation.* Washington: Congressional Research Service, Library of Congress, Feb. 10, 1978.

Bross, Irwin D, and Natarajan, H. Genetic Damage from Diagnostic Radiation. *Journal of the American Medical Association* 237:2399-2401, 1977.

Brown, Michael. *Laying Waste.* New York: Pocket Books, 1982.

Browne, Malcolm W. The Disposal Alternatives. *The New York Times Magazine,* p. 24, May 11, 1986.

Bruland, W, et al. *Radioecological Assessment of the Wyhl Nuclear Power Plant* (the "Heidelberg Report"). Heidelberg, West Germany: Department of Environmental Protection of the University of Heidelberg, rev. 1979.

Cahan, Vicky, and Harris, Catherine. VDT Safety: There May Finally Be Some Action. *Business Week,* p. 68, July 8, 1985.

Caldicott, Helen. *Nuclear Madness: What You Can Do!* New York: Bantam Books, 1979.

Carolina Nuclear Plant in Big Radioactive Leak. *The Neiv York Times,* Sept. 4, 1984.

Carothers, Andre. The Arms Race Invades Space. *New Age,* p. 10, Jan. 1983.

Carp, H, and Janoff, A. *American Review of Respiratory Disease* 118: 617-621, 1978.

Carter, Melvin, and Moghissi, Alan. Three Decades of Nuclear Testing. *Health Physics* 33:55-70, 1977.

Challem, Jack, and Lewin, Renate. Electromagnetic Radiation: A Growing Hazard. *Let's Live,* pp. 44-47, April 1986.

Chiles, James. Learning to Live with Plutonium. *Science Digest,* p. 49, July 1984.

Cummings, Judith. Reactor License Voted, But Coast Fight Goes On. *The New York Times,* p. 7, Sept. 9, 1984.

Cummings, Judith. Three Decades After Bomb Test, Utah Sheep Ranchers Feel Little Bitterness. *The New York Times,* p. 18, Aug. 15, 1982.

Curry, Bill. A-Test Officials Feared Outcry After Health Study. *The Washington Post,* April 14, 1979.

Curtis, Richard, and Hogan, Elizabeth. *Nuclear Lessons.* Harrisburg, PA: Stackpole Books, 1980.

Deacon, Kathy, VDT Hazards. *The Village Voice,* p. 54, Dec. 13, 1983.

Diamond, Stuart. Chernobyl Causing Big Revisions in Global Nuclear Power Policies. *The New York Times,* p. 1, Oct. 27, 1986.

Diamond, Stuart. Nuclear Plants Lose Cost Edge. *The New York Times, p.* 30, Sept. 3, 1984.

Diamond, Stuart. Problems Delay Three Mile Island Work. *The New York Times,* p. A12, July 25, 1984.

Divers Off Belgian Coast Recover One-Third of Radioactive Cargo. *The New York Times,* p. 4, Sept. 15, 1984.

Don't Sit Too Close to the TV: VDTICRT's and Radiation (pamphlet). Bronx, New York: Cold Type Organizing Committee, 1980.

Douglis, Carole. Stone Sentry. *Omni,* p. 62, Nov. 1985.

Dunlap, David. Shipments of Radioactive Waste Through City Are Planned Again. *The New York Times,* p. 1A, April 24, 1984.

Evanoff, Mark. Plutonium for Weapons. *Not Man Apart,* Oct. 1982.

Fadiman, Anna. The Downwind People: A Thousand Americans Sue for Damage Brought on by Atomic Fallout. *Life,* pp. 32-40, June 1980.

The Fire Unleashed. *Closeup* (ABC-TV Show). Show #125, June 6, 1985.

Fishbowl Task: Deciding How to Dispose of Nuclear Waste. *The New York Times,* p. A14, May 16, 1983.

5 Atomic Plants Ordered Shut to Inspect Pipes for Cracks. *The New York Times,* p. Bll, July 15, 1983.

5 to 6 Operators Fail to Pass Test at Largest Atomic Plant. *The New York Times,* Oct. 7, 1984.

Ford, Daniel. *The Cult of the Atom: The Secret Papers of the Atomic Energy Commission.* New York: Simon and Schuster, 1984.

Franke, B, Kruger, E, and Steinhilber-Schwab, B. *Radiation Exposure to the Public from Radioactive Emissions of Nuclear Power Stations.* Heidelberg, West Germany: Institute for Energy and Environmental Research (IFEU). Translated by the NRC, 1980.

Freeman, Leslie J. *Nuclear Witnesses: Insiders Speak Out.* New York: WW Norton & Co., 1982.

Gerber, G., and Altman, R. *Radiation Biochemistry.* New York: Academic Press, 1970.

Gofman, John W. The Cancer Hazard from Inhaled Plutonium. *Congressional Record* 121:S14610-S14616, 1975.

Gofman, John W. Estimated Production of Human Lung Cancers by Plutonium from Worldwide Fallout. *Congressional Record* 121:S14616-S14619, 1975.

Gofman, John W. Low Dose Radiation, Chromosomes and Cancer. Paper presented at the Institute for Electric, Electronic Engineers (IEEE) Nuclear Science Symposium, San Francisco: Oct. 24, 1969.

Gofman, John W. The Plutonium Controversy. *Journal of the American Medical Association* 236:284, 1976.

Gofman, John W. *Radiation and Human Health: A Comprehensive Investigation of the Evidence Relating Low-Level Radiation to Cancer and Other Diseases.* San Francisco: Sierra Club Books, 1981.

Gofman, John W, and O'Connor, Egan. *X-Rays: Health Effects of Common Exams.* San Francisco: Sierra Club Books, 1985.

Gofman, John W, and Tamplin, Arthur R. *Poisoned Power: The Case Against Nuclear Power Plants Before and After Three Mile Island.* Emmaus, PA: Rodale Press, 1979.

Gordon, Joshua. *Nuclear Power Safety Report 1979-1985.* Washington: Public Citizen—Critical Mass Energy Project, May 1986.

Graham, CL. Radiation Dose Rates from Various Smoke Detectors. *Fire Journal, p.* 109, July 1978.

Grieves, Robert T. A 1.6 Billion Nuclear Fiasco—Mismanagement and Safety Lapses Endanger an Ohio Plant. *Time Magazine,* p. 96, Oct. 31, 1983.

Grosch, DS, and Hopwood, LE. *Biological Effects of Radiation.* New York: Academic Press, 1979.

Grossman, Karl. *Cover-Up: What You Are Not Supposed to Know About Nuclear Power.* Sag Harbor, NY: Permanent Press, 1982.

Gyorgy, Anna, and Friends. *No Nukes: Everyone's Guide to Nuclear Power.* Boston: South End Press, 1979.

Hall, Eric. *Radiation and Life.* Oxford: Pergamon Press, 1976.

Harley, NH, et al. Polonium-210 in Tobacco. Paper presented at the Symposium on Public Health Aspects of Radioactivity in Consumer Products in Atlanta, GA: Feb. 2-4, 1977.

Hartmann, Thomas. Computers and Your Health: Is Radiation from VDTs a Clear and Present Danger? *East West Journal,* pp. 18-21, May 1984.

Hayes, Thomas. Diablo Canyon Reactor Starts Up Amid Protests and Industry Praise. *The New York Times,* p. 1, April 30, 1984.

Health Hazards of Electrical Wiring Cited. *Vegetarian Times, p.* 12, May 1984.

Jacobs, Paul. Precautions Are Being Taken by Those Who Know. *The Atlantic Monthly,* Feb. 1971.

Jaffe, Susan. No Nuke Is An Island: TMI Four Years Later. *Whole Life Times,* p. 13, March 1983.

Johnson, Carl J. Investigation of Health Effects in Populations Living Near a Nuclear Installation (letter). *American Journal of Public Health* 5:598-600,1973.

Journal Publishes Disputed Study on Cancer Increase in Mormons. *The New York Times,* Jan. 13, 1984.

Katheren, Ronald. What Is Occupational Exposure? *Health Physics* 39(2):141, 1980.

Katz, Arthur. *Life After Nuclear War.* Cambridge, MA: Ballinger Publishing Co., 1982.

Kaufer, Scott. The Air Pollution You Can't See. *New Times,* pp. 29-37, 60, March 6, 1978.

Kennan, George F. *Nuclear Delusion: Soviet-American Relations in the Atomic Age.* New York: Random House, Pantheon Books, 1982.

Key, MM, et al., Eds. *Occupational Diseases: A Guide to Their Recognition.* National Institute for Occupational Safety and Health (NIOSH), 1977.

Koen, Susan, et al. *A Handbook for Women on the Nuclear Mentality.* Norwich, VT: Wand, 1980.

Larsen, RD, and Oldham, RD. Plutonium in Drinking Water: Effects of Chlorine on Its Maximum Permissible Concentration. *Science* 201:1008-1009, 1978.

The League of Women Voters Education Fund. *The Nuclear Waste Primer: A Handbook for Citizens.* New York: Nick Lyons Books, 1985.

The Lethal Shuttle (editorial). *The Nation,* p. 1, Feb. 22, 1986.

Lipschutz, Ronnie D. *Radioactive Waste: Politics, Technology and Risk: A Report of the Union of Concerned Scientists.* Cambridge, MA: Ballinger Publishing Co., 1980.

Litwak, Mark. Would You Rather Fight than Switch? *Whole Life Times,* p. 11, April/May 1985.

Loeb, Paul. *Nuclear Culture.* Philadelphia: New Society Publishers, 1986.

Lowder, W, and Condon, W. Measurement of the Exposure of Human Populations to Environmental Radiation. *Nature* 206:658-662, 1965.

A *"Low-Level" Nuclear Waste Primer.* New York: Sierra Club Radioactive Waste Campaign, 1985.

Luck, Werner, and Nav, Heinz. Exposure of the Fetus, Neonate, and Nursed Infant to Nicotine and Cotinine for Maternal Smoking. *New England Journal of Medicine* 311:672, 1984.

McKinley, C. *Unacceptable Risk.* New York: Bantam Books, 1976.

Malcomson, Scott L. High Anxiety: A Journey Through the Atomic Hinterland. *The Village Voice,* pp. 29-31, Nov. 5, 1985.

Martell, Edward A. Radioactivity of Tobacco Trichomes and Insoluble Smoke Particles. *Nature* 247:215-217, 1974.

Martell, Edward A. Tobacco Radioactivity and Cancer in Smokers. *American Scientist* 63:404, 1975.

Marx, Jean L. Low-Level Radiation: Just How Bad Is It? *Science* 204:160-164, 1979.

Mayers, CP, and Habeshaw, JA. A Non-Thermal Effect of Micro-Wave Radiation as a Potential Hazard to Health. *International Journal of Radiation Biology* 24:449-461, 1973.

Maynard, Joyce. The Story of a Town. *The New York Times Magazine,* p. 24, May 11, 1986.

Nader, Ralph, and Abbots, John. *The Menace of Atomic Energy.* New York: WW Norton & Co., 1977.

Nader, Ralph, et al. *Who's Poisoning America?* San Francisco: Sierra Club Books, 1981.

National Council on Radiation Protection and Measurements. *Basic Radiation Protection Criteria: Recommendations.* NCRP Report #39. Washington: US Government Printing Office, 1971, 1980.

National Council on Radiation Protection and Measurements. *Radiation Exposure from Consumer Products and Miscellaneous Sources.* NCRP Report #56. Washington: US Government Printing Office, 1977.

Nero, Anthony V. The Indoor Radon Story. *Technology Review,* p. 28, Jan. 1986.

Ninth Annual Occupational Radiation Report. United States Nuclear Regulatory Commission, Oct. 1977.

Not in My Backyard: Low-Level Radioactive Waste and Health. *The Harvard Medical School Health Letter* 11:4-6, 1986.

Nuclear Agency Removes Investigative Chief. *The New York Times,* p. A20, Sept. 29, 1983.

Nuclear America (map), 2nd ed. New York: War Resisters League, 1979.

The Nuclear Bargain. *Newsweek,* p. 40, May 12, 1986.

Nuclear Leak Data Criticized. *The New York Times,* p. A17, April 17,1984.

The Nuclear Power Controversy. Brochure from Union of Concerned Scientists, Cambridge, MA.

Nuclear Power Plants in the United States (map) by the Atomic Industrial Forum, Bethesda, MD, 1986.

Nuclear Shipments: Accidents on the Rise. *The Guardian,* Dec. 3, 1980.

Nuclear Weapons Locations in the US (map) by the Center for Defense Information, Washington, DC, 1982.

Olson, Steve. Nuclear Undertakers. *Science,* pp. 50-59, Sept. 1984.

Ott, John. *International Journal for Biosocial Research,* July 1985.

Ott, John. *Light, Radiation, and You.* Old Greenwich, CT: Devin-Adair Co., 1982.

Plea to Shut Down Reactor Is Denied. *The New York Times,* p. B2, Sept. 29, 1983.

Pollard, Robert, Ed. *The Nugget File.* Cambridge, MA: Union of Concerned Scientists, 1979.

Ponte, Lowell. Radioactivity: The New-Found Danger in Cigarettes. *Reader's Digest,* pp. 123-127, March 1986.

Population Reports. *Tobacco, Hazards to Human Health and Reproduction.* Baltimore, MD: Johns Hopkins University, 1979.

Pringle, Laurence. *Radiation: Waves and Particles/Benefits and Risks.* Hillside, NJ: Enslow Publishers, 1983.

Radford, EP, and Hunt, VR. Polonium-210: A Volatile Radioelement in Cigarettes. *Science* 143(3603):247-249, 1964.

Radiation: A Fact of Life, printed and distributed by the American Nuclear Society with the permission of the International Atomic Energy Agency, 1980.

Radioactive Waste: Buried Forever? A Profile of Commercial Radioactive Landfills. *A Radioactive Waste Campaign Fact Sheet.* New York: Radioactive Waste Campaign, 1985.

Radon in Houses. *Harvard Medical School Health Letter* 11(7):1-3, 1986.

Radon May Endanger 8 Million Homes. *The New York Times,* November 15, 1985.

Rappoport, Roger. *The Great American Bomb Machine.* New York: E.P. Dut-ton, 1971.

Reactor Science and Technology vol 2(3), Oct. 1952. A 1982 declassified version of this document is available from Oak Ridge Technical Library, Department of Energy, P.O. Box E, Oak Ridge, TN 37830.

The Reactor That May Break the Bank. *The New York Times,* Oct. 30, 1982.

Rules Approved on Low-Level Radioactive Waste. *The New York Times,* p. 23, Oct. 24, 1982.

Saiter, Susan. Shipments of Nuclear Waste Are Increasing Dramatically. *The New York Times,* p. 1, August 21, 1983.

Samuels, Mike, and Bennett, Hal Zina. *Well Body, Well Earth: The Sierra Club Environmental Health Sourcebook,* San Francisco: Sierra Club Books, 1983.

Schell, Jonathan. *The Fate of the Earth.* New York: Alfred A. Knopf, 1982.

Schmidt, William. Federal Judge Says US Deceived Him in '56 Fallout Case. *The New York Times,* Aug. 4, 1982.

Scoville, Herbert. MX: *Prescription for Disaster.* Cambridge, MA: MIT Press, 1981.

Segal, A, and Reed, R. Human Exposure to External Background Radiation. *Archives of Environmental Health* 9:492-499, 1964.

Sellafield to Fight Radioactive Leak Charges in Court. *The Times* (London), p. 5, Oct. 5, 1984.

Shapiro, Fred C. *Radwaste.* New York: Random House, 1981.

Six Northeast States Discuss Radioactive Waste Disposal. *The New York Times,* p. A36, April 15, 1984.

Slesin, Louis. People Are Antennas, Too: The Biology of the Electromagnetic Spectrum. *Whole Earth Review,* pp. 48-54, Spring 1986.

Sloyan, Patrick J. Nuclear Time Bomb. *Newsday,* May 20, 1986.

Smay, VE. Radon Exclusive. *Popular Science,* p. 77, Nov. 1985.

Smith, RJ. Atom Bomb Tests Leave Infamous Legacy. *Science* 218:269, Oct. 15, 1982.

Smith, RJ. Scientists Implicated in Atom Test Deception. *Science* 218:545-547, 1982.

Spengler, John. Indoor Air Pollution. *Harvard Medical School Health Letter* 9:3-5, 1984.

Stern, Phyllis. Three Mile Island on Wheels. *Whole Life Times,* p. 1, Oct. 1984.

Sternglass, Ernest J. Cancer Mortality Changes around Nuclear Facilities in Connecticut. Testimony presented in *Radiation Standards and Public Health—Proceedings of a Second Congressional Seminar on Low-Level Ionizing Radiation.* Washington, DC: Congressional Research Service, Library of Congress, Feb. 10, 1978.

Sternglass, Ernest J. The Death of All Children. *Esquire Magazine,* la-Id, Sept. 1969.

Sternglass, Ernest J. The Role of Indirect Radiation Effects on Cell Membranes in the Immune Response. *Radiation and the Immune Process.* Proceedings of the 1974 Hanford Radiobiology Symposium, Division of Technical Information, ERDA, Oak Ridge, TN (Cong. 740930).

Sternglass, Ernest J. *Secret Fallout: Low-Level Radiation from Hiroshima to Three Mile Island.* New York: McGraw-Hill, 1981.

Sun, Marjorie. Static at EPA Over Broadcast Transmitters. *Science* 225:32-33, 1984.

13 A-Plant Guards Ousted. *The New York Times,* p. A13, Aug. 27, 1983.

Town Ties Cancer Deaths to Silo Containing Atomic-Bomb Waste. *The New York Times,* May 8, 1979.

Tucker, Kitty, and Haas, Lesley. *Nuclear Production—U.S.A.* (map). Washington: Health and Energy Institute, 1986.

A Twenty Year Review of Medical Findings in a Marshalese Population Accidentally Exposed to Radiation Fallout, pp. 39-62. Upton, NY: Brookhaven National Laboratory, 1975.

T.V. A. Lays Off Hundreds Over Nuclear Plant Safety. *The New York Times,* p. A8, Jan. 24, 1984.

United Nations Scientific Committee on the Effects of Atomic Radiation (UNSCEAR). *Sources and Effects of Ionizing Radiation,* 1982.

U.S. Citizens Near Bomb Plant: Concerns in Washington State Prompt Radiation Exams. *The New York Times,* Dec. 16, 1985.

U.S. Checking Nuclear Plants on Faulty Parts. *The New York Times,* p. 1, Jan. 23, 1983.

U.S. Expects 11 Nuclear Plants Will Need Changes. *The New York Times,* p. A17, Oct. 4, 1982.

US Office on Smoking and Health. *Smoking and Health: A Report of the Surgeon General.* US Dept. of Health, Education and Welfare, Public Health Service. Washington: US Government Printing Office, 1979.

US Revives Space Nuclear Power. *High Technology* 4(8):15-19, 1984.

VDT's—A New Social Disease? *The Harvard Medical School Health Letter,* vol. 8, 1983.

Wald, Matthew L. By 2005, Nuclear Unit Sees 50-50 Chance of Meltdown. *The New York Times, p.* A16, April 17, 1986.

Wald, Matthew L. Northwest Plutonium Plant Had Big Radioactive Emissions in 40's and 50's. *The New York Times,* p. A20, Oct. 24, 1986.

Wald, Matthew L. Report Assails Safety of Nuclear Waste Storage at Carolina Plant. *The New York Times, p.* A11, July 24, 1986.

Wasserman, Harvey, and Solomon, Norman. Every American Put at Risk—New Light on the Dangers of Low-Level Radiation. *The Nation, p.* 1, Jan. 1-8, 1983.

Wasserman, Harvey, and Solomon, Norman. *Killing Our Own: The Disaster of America's Experience with Atomic Radiation.* New York: Dell Publishing Co., 1982.

The Waste Paper. Summer 1985.

Weast, Robert C, Ed. *Handbook of Chemistry and Physics,* 63rd edition. Boca Raton, FL: CRC Press, 1982.

Williams, Roger. *Nutrition Against Disease.* Marshfield, MA: Pitman Publishing, 1971.

World Health Organization VDT Working Group. *VDT News,* pp. 12-13, Jan./Feb. 1986.

Wynder, EL, and Hoffman, D. *Tobacco and Tobacco Smoke; Studies in Experimental Carcinogenesis.* New York: Academic Press, 1967.

X-Rays of Welds Altered at A-Plant. *The New York Times,* p. B2, Oct. 19,1982.

Yavelow,], et al. The Bowman-Birk Soybean Protease Inhibitor as Anticar-cinogen. *Cancer Research* 43:2454-2459, 1983.

Chapter Two

Adams, C, and Bonnel, J. Administration of Stable Iodide as a Means of Reducing Thyroid Irradiation Resulting From Inhalation of Radioactive Iodine. *Health Physics* 7:127, 1962.

Advisory Committee on the Biological Effects of Ionizing Radiation (BEIR), 1972 Report. *The Effects on Population of Exposure to Low Levels of Ionizing Radiation.* Washington: National Academy Press, 1972.

Akizuki, Tatsuichiro. *Nagasaki 1945.* London: Quartet Books, 1981.

Alpert, ME, et al. Chemical and Radionuclide Food Contamination. *MMS Information Corp.,* 1973.

Ames, Bruce N. Dietary Carcinogens and Anticarcinogens: Oxygen Radicals and Degenerative Diseases. *Science* 23:1256-1263, 1983.

Anomalies Occurring in Children Exposed in Utero to the Atomic Bomb in Hiroshima. *Pediatrics* 10:687-692, 1952.

Arena, Victor. *Ionizing Radiation and Life.* St. Louis: CV Mosby, 1971.

Asimov, Isaac, and Dobzhansky, Theodosius. *The Genetic Effects of Radiation.* US AEC Division of Technical Information, pp. 26-38, 1966.

Austin, D, Snyder H, et al. Malignant Melanoma Among Employees of Lawrence Livermore National Laboratory. *Lancet* 2:712-713, 1981.

Bailey, Herbert. *E: The Essential Vitamin.* New York: Bantam Books, 1983.

Begley, Sharon. The 20th-century Plague. *Newsweek,* pp. 36-37, May 12, 1986.

Beirwaltes, WH, et al. Radioactive Iodine Concentration in the Fetal Human Thyroid Gland. *Journal of the American Medical Association* 173:1185-1902,1960.

Berger, John J. *Nuclear Power: The Unviable Option.* Palo Alto, CA: Ramparts Press, 1976.

Bertell, Rosalie. *No Immediate Danger?—Prognosis for a Radioactive Earth.* Summertown, TN: The Book Publishing Co., 1986.

Bertell, Rosalie. Radiation Exposure and Human Species Survival. *Environmental Health Review,* pp. 43-52, June 1981.

Bertell, Rosalie. X-Ray Exposure and Premature Aging. *Journal of Surgical Oncology* 9:379-391, 1977.

Bowen, VT. Transuranic Elements and Nuclear Wastes. *Oceanus* 18:48, 1974.

Brickenfeld, Dick. A New German Study Challenges the NRC's Assurances. *The Washington Post,* Nov. 11, 1979.

Bross, Irwin D. Hazards to Persons Exposed to Ionizing Radiation (and to their children) from Dosage Currently Permitted by the Nuclear Regulatory Commission, pp. 913-950 in *Effect of Radiation on Human Health: Health Effects of Ionizing Radiation,* vol. 1, Hearings before the Subcommittee on Health and the Environment ot the Committee on Interstate and Foreign Commerce of the U.S. House of Representatives. Serial No. 95-179. Washington: Jan./Feb. 1978.

Bross, Irwin D, et al. A Dosage Response Curve for the One Rad Range: Adult Risks from Diagnostic Radiation. *American journal of Public Health* 69:130-136, 1979.

Bross, Irwin D, and Natarajan, H. Genetic Damage from Diagnostic Radiation, *journal of the American Medical Association* 237:2399-2401, 1977.

Bross, Irwin D, and Natarajan, H. Leukemia from Low-Level Radiation: Identification of Susceptible Children. *New England journal of Medicine* 287:107-110, 1972.

Bross, Irwin D, and Natarajan, H. Risk of Leukemia in Susceptible Children Exposed to Pre-Conception, in Utero, and Postnatal Radiation. *Preventive Medicine* 3:361-369, 1974.

Bruland, W, et al. *Radioecological Assessment of the Wyhl Nuclear Power Plant* (the "Heidelberg Report"). Heidelberg, West Germany: Department of Environmental Protection of the University of Heidelberg, rev. 1979.

Caldicott, Helen. *Nuclear Madness: What You Can Do!* New York: Bantam Books, 1979.

Cannon, PR. Antibodies and Protein Reserves, *journal of Immunology* 44:107, 1942.

Cline, J. Effect of Nutrient Potassium on the Uptake of Cesium 137 and Potassium and on Discrimination Factor. *Nature* 193:1302-1303, 1962.

DeGroot, Morris. Statistical Studies of the Effect of Low-Level Radiation from Nuclear Reactors on Human Health, in Neyman, J, Ed., *Proceedings of Sixth Berkeley Symposium on Mathematical Statistics and Probability.* Berkeley: University of California Press, 1971.

Doll, Richard, and Peto, Richard. The Causes of Cancer: Quantitative Estimates of Avoidable Risks of Cancer in the United States Today, *journal of the National Cancer Institute* 66:1193-1208, 1981.

Elam, S. The Nuclear Radiation/SAT Connection. *Phi Delta Kappan* 61:184-187, 1979.

Fabrikant, Jacob. Estimation of Risk of Cancer Induction in Populations Exposed to Low-Level Radiation. *Investigative Radiology* 17:342-349, 1982.

Franke, B, Kruger, E, and Steinhilber-Schwab, B. *Radiation Exposure to the Public from Radioactive Emissions of Nuclear Power Stations.* Heidelberg, West Germany: Institute for Energy and Environmental Research (IFEU). Translated into English by the NRC, 1980.

Freeman, Leslie J. *Nuclear Witnesses: Insiders Speak Out.* New York: WW Norton & Co., 1982.

Gerber, et al. Effects of Localized Radiotherapy upon the Cellular Immune Response. *Radiation Research* 85:390-398, 1981.

Gerber, G, and Altman, R. *Radiation Biochemistry.* New York: Academic Press, 1970.

Gessell, TF, and Prichard, HM. The Technologically Enhanced Natural Radiation Environment. *Health Physics* 28:361-366, 1975.

Gofman, John W. *Cancer Hazards from Low-Dose Radiation.* United States of America Nuclear Regulatory Commission before the Hearing Board, 1977.

Gofman, John W. *Radiation and Human Health: A Comprehensive Investigation of the Evidence Relating Low-Level Radiation to Cancer and Other Diseases.* San Francisco: Sierra Club Books, 1981.

Gofman, John W, and Tamplin, Arthur R. *Poisoned Power: The Case Against Nuclear Power Plants Before and After Three Mile Island.* Emmaus, PA: Rodale Press, 1979.

Grosch, DS, and Hopwood, LE. *Biological Effects of Radiation.* New York: Academic Press, 1979.

Harman, Denham. Free Radical Theory of Aging. *Journal of Gerontology* 23(4):476-482, 1968.

Harman, Denham. Free Radical Theory of Aging—Nutritional Implications. *Age* 1(4):145-152, 1978.

Harman, Denham. Role of Free Radicals in Mutation, Cancer, Aging and the Maintenance of Life. *Radiation Research* 16:753-764, 1962.

Harman, Denham, Hendricks, S, and Eddy, D. Free Radical Theory of Aging: Effects of Free-Radical-Reaction Inhibitors on the Immune Response. *Journal of American Geriatrics Society,* 25:400-407, 1977.

Harvey, Elizabeth, et al. Prenatal X-Ray Exposure and Childhood Cancer in Twins. *New England Journal of Medicine* 312:541-545, 1985.

Heal Thyself. *East West Journal,* p. 36, March 1983.

Holden, Constance. Low-Level Radiation: A High-Level Concern. *Science* 204:155-158, 1979.

Johnson, Carl J. Cancer Incidence in an Area of Radioactive Fallout Downwind from the Nevada Test Site. *Journal of the American Medical Association* 251:230-236, 1984.

Journal Publishes Disputed Study on Cancer Increase in Mormons. *The Neiv York Times,* Jan. 13, 1984.

Kensler, TW. Inhibition of Tumor Promotion by a Biomimetic Superoxide Dismutase. *Science* 221:75-77, 1983.

Kushi, Michio, and Jack, Alex. *The Cancer-Prevention Diet: Michio Kushi's Nutritional Blueprint for the Relief and Prevention of Disease.* New York: St. Martin's Press, 1983.

Lenihan, JM, Ed. *Strontium Metabolism.* New York: Academic Press, 1967.

Longevity and Causes of Death from Irradiation of Physicians. *Journal of the American Medical Association* 162:464, 1956.

Lyon, Joseph L, et al. Childhood Leukemias Associated with Fallout from Nuclear Testing. *New England Journal of Medicine* 300:397-402, 1979.

Mancuso, Thomas F, Stewart, Alice M, and Kneale, George W. Radiation Exposures of Hanford Workers Dying from Cancer and Other Causes. *Health Physics* 33:369-384, 1977.

Mann, John. *Secret of Life Extension.* New York: Bantam Books, 1980.

Martell, Edward A. *Cesium-137 From the Environment to Man: Metabolism and Dose.* NCRP Report #52. Washington: National Council on Radiation Protection and Measurements, 1977.

Monson, RR, and MacMahon, B. Prenatal X-Ray Exposure and Cancer in Children. *Radiation Carcinogenesis: Epidemiology and Biological Significance,* pp. 97-105, 1984.

Morgan, Karl J. Cancer and Low-Level Ionizing Radiation. *Bulletin of the Atomic Scientists,* p. 30, Sept. 1978.

Najarian, Thomas, and Colton, Theodore. Mortality from Leukemia and Cancer in Shipyard Workers. *Lancet* 147(l):1018-1020, May 13, 1978.

National Council on Radiation Protection and Measurements. *Basic Radiation Protection Criteria: Recommendations.* NCRP Report #39. Washington: US Government Printing Office, 1971, 1980.

National Council on Radiation Protection and Measurements. *NCRP Review of NCRP Radiation Dose Limit for Embryo and Fetus in Occupationally Exposed Women.* NCRP Report #53. Washington: US Government Printing Office, 1977.

National Council on Radiation Protection and Measurements. *Radiation Exposure from Consumer Products and Miscellaneous Sources.* NCRP Report #56. Washington: US Government Printing Office, 1977.

National Research Council. *Diet, Nutrition, and Cancer.* Washington: National Academy Press, 1982.

New A-Bomb Studies Alter Radiation Estimates. *Science* May 22, 1981.

Norman, Colin. Hazy Picture of Chernobyl Engineering. *Science* 232:1331-1332, 1986.

Norwood, Christopher. *Terata.* San Francisco: Mother Jones, 1985.

Ohsawa, George. *Cancer and the Philosophy of the Far East.* Oroville, CA: George Ohsawa Macrobiotic Foundation, 1981.

Pageau, R, et al. Systemic Protection Against Radiation: Effect of an Elemental Diet on Hematopoietic and Immunological Systems in the Rat. *Radiation Research* 62:357-363, 1975.

Palmer, R, et al. Effect of Calcium Deposition of Strontium 90 and Calcium 45 in Rats. *Science* 127:1505, 1958.

Part II, Title 10, Parts 19 et al., Standards for Protection Against Radiation. *Code of Federal Regulations,* January 1986.

Passwater, Richard, and Cranton, Elmer M. *Trace Elements, Hair Analysis and Nutrition.* New Canaan, CT: Keats, 1983.

Pauling, Linus, How Dangerous Is Radioactive Fallout? *Foreign Policy Bulletin,* June 15, 1957.

Pauling, Linus. *No More War.* New York: Dodd, Mead & Co., 1958.

Pearson, Durk, and Shaw, Sandy. *Life Extension.* New York: Warner Books, 1983.

Petkau, Abram. Effect of 22Na+ on a Phospholipid Membrane. *Health Physics* 22:239-244, 1972.

Pfeiffer, Carl C. *Mental and Elemental Nutrients.* New Canaan, CT: Keats, 1975.

Potassium Iodide as a Thyroid-Blocking Agent in a Radiation Emergency: Draft Recommendations on Use. *Federal Register* 46:30199, 1981.

Price, KR. *A Critical Revieiv of Biological Accumulation, Discrimination, and Uptake of Radionuclides Important to Waste Management Practices, 1943-1971,* 1971.

Priest, ND. The Distribution of Plutonium-241 in Rodents. *International journal of Radiation Biology* 31:59-78, 1977.

Pryor, WA. Free Radical Reactions and their Importance in Biological Systems. *Federation of American Societies for Experimental Biology. Federation Proceedings.* 22:1982, 1973.

Rafla, S, et al. Changes in Cell-Mediated Immunity in Patients Undergoing Radiotherapy. *Cancer* 41:1076-1086, 1978.

Rous, P. The Influence of Diet on Transplanted and Spontaneous Mouse Tumors. *Journal of Experimental Medicine* 20:433, 1914.

Saxena, K, et al. Minimal Dosage of Iodide Required to Suppress Uptake of Iodine-131 by Normal Thyroid. 138:430, 1962.

Shortening of Life by Chronic Irradiation: The Experimental Facts. *Nature* 179:456, 1957.

Shutdown: Nuclear Power on Trial. Transcript of proceedings in Federal Court. Summertown, TN: The Book Publishing Co., 1979.

Simone, Charles. *Cancer and Nutrition.* New York: McGraw-Hill, 1983.

Spencer, H, et al. Effect of Magnesium on Radiostrontium Excretion in Man. *international journal of Applied Radiation and Isotopes* 18:407-415, 1967.

Steele, Karen Dorn. Radiation Exposure May Lower Fetus' IQ. *Spokesman-Review* (Spokane, WA), pp. 1-2, Jan. 5, 1986.

Sternglass, Ernest J. Can the Infants Survive? *Bulletin of Atomic Scientists* 25:26, 1969.

Sternglass, Ernest J. Cancer: Relation of Prenatal Radiation to Development of the Disease in Childhood. *Science* 140:1102-1104, 1963.

Sternglass, Ernest J. The Death of All Children. *Esquire Magazine,* pp. la-Id, Sept. 1969.

Sternglass, Ernest J. The Implications of Chernobyl for Human Health. *International Journal of Biosocial Research* 8:7-36, 1986.

Sternglass, Ernest J. Infant Mortality and Nuclear Tests. *Bulletin of Atomic Scientists,* 25:18-20, April 1969.

Sternglass, Ernest J. *Secret Fallout: Low-Level Radiation from Hiroshima to Three Mile Island.* New York: McGraw-Hill, 1981.

Sternglass, Ernest J, and Bell, S. Fallout and SAT Scores: Evidence for Cognitive Damage During Early Infancy. *Phi Delta Kappan* 65:539-545, 1983.

Sternglass, Ernest J, and Bell, S. Fallout and the Decline of Scholastic Aptitude Scores. Paper presented at the Annual Meeting of the American Psychological Association, New York, NY: Sept. 3, 1979.

Stewart, Alice M. Delayed Effects of A-Bomb Radiation: A Review of Recent Mortality Rates and Risk Estimates for Five Year Survivors. *Journal of Epidemiology and Community Health* 36(2):80-86, 1982.

Stewart, Alice M, and Hewitt, David. Leukemia Incidence in Children in Relation to Radiation Exposure in Early Life, in *1965 Current Topics in Radiation Research,* Ebert, Michael and Howard, Alma, Eds. North Holland Publishing Co., 1965.

Stewart, Alice M, and Kneale, George W. Radiation Dose Effects in Relation to Obstetric X-Rays and Childhood Cancers. *Lancet,* pp. 1185-1188, June 6, 1970.

Stewart, Alice M, and Kneale, George W. Radiation Exposures of Workers Dying from Cancer and Other Causes (letter). *Health Physics* 36(1):87, 1977.

Stewart, Alice M, Webb, J, and Hewitt, David. A Survey of Childhood Malignancies. *British Medical Journal* 1:1495-1508, 1958.

Storer, J, et al. Life Shortening in RFM and BALB Mice as a Function of Radiation Quality Dose and Dose Rate. *Radiation Research* 78:122-161, 1979.

Szentkuti, L, and Giese, W. Studies on the Bending Capacity of the Skeletal Muscle Constituents for Cesium-137. **Health** *Physics* 26:343-347, 1974.

Tannenbaum, A. The Initiation and Growth of Tumors: Effects of Underfeeding. *American journal of Cancer* 38:335, 1940.

Trabelka, J, and Garten, C. *Advances in Radiation Biology.* New York: Academic Press, 1983.

Tu, Anthony T. *Survey of Contemporary Toxicology.* New York: John Wiley & Sons, 1982.

Waldron, C, Correll, L, Soignier, M, and Puck, T. Measurement of Low Levels of X-Ray Mutagenesis in Relation to Human Disease. *National*

Academy of Sciences of the United States of America. Proceedings. Biological Sciences 83:4839-4843, 1986.

Wasserman, H, and Comar, C. Effect of Dietary Calcium and Phosphorus Levels on Body Burden of Ingested Radiostrontium. *Proceedings. Society for Experimental Biology and Medicine* 103:124, 1960.

Wasserman, Harvey, and Solomon, Norman. Every American Put At Risk—New Light on the Dangers of Low-Level Radiation. *The Nation,* p. 1, Jan. 1-8, 1983.

Wasserman, Harvey, and Solomon, Norman. *Killing Our Own: The Disaster of America's Experience with Atomic Radiation.* New York: Dell Publishing Co., 1982.

Williams, Roger. *Nutrition Against Disease.* Marshfield, MA: Pitman Publishing, 1971.

Wood, L, et al. *Your Thyroid: A Home Reference.* Boston: Houghton Mifflin, 1982.

Chapter Three

Adams, C, and Bonnel, J. Administration of Stable Iodide as a Means of Reducing Thyroid Irradiation Resulting From Inhalation of Radioactive Iodine. *Health Physics* 7:127, 1962.

Akizuki, Tatsuichiro. *Nagasaki 1945.* London: Quartet Books, 1981.

Alpert, ME, et al. Chemical and Radionuclide Food Contamination. *MMS Information Corp.,* 1973.

Ames, Bruce N. Dietary Carcinogens and Anticarcinogens: Oxygen Radicals and Degenerative Diseases. *Science* 23:1256-1263, 1983.

Arena, Victor. *Ionizing Radiation and Life.* St. Louis: CV Mosby, 1971.

Bailey, Herbert. *E: The Essential Vitamin.* New York: Bantam Books, 1983.

Beirwaltes, WH, et al. Radioactive Iodine Concentration in the Fetal Human Thyroid Gland, *journal of the American Medical Association* 173:1185-1902,1960.

Berger, John J. *Nuclear Power: The Unviable Option.* Palo Alto, CA: Ramparts Press, 1976.

Bruland, W, et al. *Radioecological Assessment of the Wyhl Nuclear Power Plant* (the "Heidelberg Report"). Heidelberg, West Germany:

Department of Environmental Protection of the University of Heidelberg, rev. 1979.

Caldicott, Helen. *Nuclear Madness: What You Can Do!* New York: Bantam Books, 1979.

Cannon, PR. Antibodies and Protein Reserves, *journal of Immunology* 44:107, 1942.

Cline, J. Effect of Nutrient Potassium on the Uptake of Cesium 137 and Potassium and on Discrimination Factor. *Nature* 193:1302-1303, 1962.

Falck, F. and Keren, DF. Protein Excretion Patterns in Cadmium-Exposed Individuals. *Archives of Environmental Health* 39(2):69-73, 1984.

Grosch, DS and Hopwood, LE. *Biological Effects of Radiation.* New York: Academic Press, 1979.

Heal Thyself. *East West Journal,* p. 36, March 1983.

International Journal of Applied Radiation and Isotopes 18:407-415, 1967.

Kushi, Michio. *The Book of Macrobiotics: The Universal Way of Health and Happiness.* Tokyo: Japan Publications, 1977.

Kushi, Michio, and Jack, Alex. *The Cancer-Prevention Diet: Michio Kushi's Nutritional Blueprint for the Relief and Prevention of Disease.* New York: St. Martin's Press, 1983.

Lenihan, JM, Ed. *Strontium Metabolism.* New York: Academic Press, 1967.

Martell, Edward A. *Cesium-137 From the Environment to Man: Metabolism and Dose.* NCRP Report #52. Washington: National Council on Radiation Protection and Measurements, 1977.

National Council on Radiation Protection and Measurements. *Protection of the Thyroid Gland in the Event of Release of Radioiodine.* NCRP Report #55. Washington: US Government Printing Office, Aug. 1977.

National Research Council. *Diet, Nutrition, and Cancer.* Washington: National Academy Press, 1982.

Ohsawa, George. *Cancer and the Philosophy of the Far East.* Oroville, CA: George Ohsawa Macrobiotic Foundation, 1981.

Pageau, R, et al. Systemic Protection Against Radiation: Effect of an Elemental Diet on Hematopoietic and Immunological Systems in the Rat. *Radiation Research* 62:357-363, 1975.

Palmer, R, et al. Effect of Calcium Deposition of Strontium 90 and Calcium 45 in Rats. *Science* 127:1505, 1958.

Passwater, Richard, and Cranton, Elmer. *Trace Elements, Hair Analysis and Nutrition.* New Canaan, CT: Keats, 1983.

Pfeiffer, Carl C. *Mental and Elemental Nutrients.* New Canaan, CT: Keats, 1975.

Potassium Iodide as a Thyroid-Blocking Agent in a Radiation Emergency: Draft Recommendations on Use. *Federal Register* 46:30199, 1981.

Price, KR. *A Critical Review of Biological Accumulation, Discrimination, and Uptake of Radionuclides Important to Waste Management Practices, 1943-1971,* 1971.

Priest, ND. The Distribution of Plutonium-241 in Rodents. *International Journal of Radiation Biology* 31:59-78.

Rous, P. The Influence of Diet on Transplanted and Spontaneous Mouse Tumors. *Journal of Experimental Medicine* 20:433, 1914.

Saxena, K, et al. Minimal Dosage of Iodide Required to Suppress Uptake of Iodine-131 by Normal Thyroid. 138:140, 1962.

Simone, Charles. *Cancer and Nutrition.* New York: McGraw-Hill, 1983.

Spencer, H, et al. Effect of Low and High Calcium Intake on Sr90 Metabolism in Adult Man. *International journal of Applied Radiation and Isotopes.* 18:605-614, 1967.

Stewart, Alice M. Delayed Effects of A-Bomb Radiation: A Review of Recent Mortality Rates and Risk Estimates for Five-Year Survivors. *Journal of Epidemiology and Community Health* 36(2):80-86, 1982.

Szentkuti, L, and Giese, W. Studies on the Bending Capacity of the Skeletal Muscle Constituents for Cesium-137. *Health Physics* 26:343-347, 1974.

Tannenbaum, A. The Initiation and Growth of Tumors: Effects of Underfeeding. *American Journal of Cancer* 38:335, 1940.

Trabelka, J, and Garten, C. *Advances in Radiation Biology.* New York: Academic Press, 1983.

Wasserman, H, and Comar, C. Effect of Dietary Calcium and Phosphorus Levels on Body Burden of Ingested Radiostrontium. *Proceedings. Society for Experimental Biology and Medicine* 103:124, 1960.

Wasserman, Harvey, and Solomon, Norman. *Killing Our Own: The Disaster of America's Experience with Atomic Radiation.* New York: Dell Publishing Co., 1982.

Williams, Roger. *Nutrition Against Disease.* Marshfield, MA: Pitman Publishing, 1971.

Wood, L, et al. *Your Thyroid: A Home Reference.* Boston: Houghton Mifflin, 1982.

Chapter Four

Aihara, H, and Ohsawa, G. *Acid and Alkaline.* Oroville, CA: George Ohsawa Macrobiotic Foundation, 1971.

Aiuti, F, et al. Effect of Thymus Factor on Human Precursor T-Lympho-cytes. *Clinical Experimental Immunology* 20:499-503, 1975.

Akizuki, Tatsuichiro. *Nagasaki 1945.* London: Quartet Books, 1981.

America's Toxic Protection Gap. Washington: Environmental Safety, 1984.

Annest, JL, et al. Chronological Trends in Blood Lead Levels Between 1976 and 1980. *New England Journal of Medicine* 308:1373-1377, 1983.

Black, David. The Plague Years, parti. *Rolling Stone,* p. 48, March 28,1985.

Black, David. The Plague Years, part2. *RollingStone,* p. 45, April 11,1985.

The Book of the Body: The Illustrated Encyclopedia of Health, Medicine and the Body. New York: Simon & Schuster, 1973.

Boulos, BM, et al. Placental Transfer of Lithium and Environmental Toxicants and Their Effects on the Newborn. *Federation of American Societies for Experimental Biology: Federation Proceedings* 32:2989, 1973.

Brown, RR. The Role of Diet in Cancer Causation. *Food Technology* 37:49, 1983.

Bruland, W, et al. *Radioecological Assessment of the Wyhl Nuclear Power Plant* (the "Heidelberg Report"). Heidelberg, West Germany: Department of Environmental Protection of the University of Heidelberg, rev. 1979.

Bryce-Smith, D, and Waldron, HA. Lead, Behaviour and Criminality. *The Ecologist* 4:90-102, 1974.

Burrow, Gerard. Caring for AIDS Patients: The Physician's Risk and Responsibility. *Canadian Medical Association Journal* 129:1181, 1983.

Cadmium: Worker and Environmental Poison. *The Public Citizen Health Research Group Health Letter* 2(3):15, 1986.

Caldicott, Helen. *Nuclear Madness: What You Can Do!* New York: Bantam Books, 1979.

Campbell, AD, et al. Immunosuppressive Consequences of Radiotherapy and Chemotherapy in Patients with Acute Lymphoblastic Leukemia. *British Medical Journal,* pp. 385-388, May 19, 1973.

Cancer Society Prepares Diet Suggested for Reduced Risks. *The New York Times,* p. A14, Sept. 30, 1983.

Chandra, RK. Cell Mediated Immunity in Genetically Obese Mice. *American Journal of Clinical Nutrition* 33:13-16, 1980.

Chandra, RK. Immunodeficiency in Under nutrition and Over nutrition. *Nutrition Reviews* 39(6):225-231, 1981.

Chandra, RK, and Scrimshaur, N. Immunocompetence in Nutritional Assessment. *American Journal of Clinical Nutrition* 33:2694-2697, 1980.

Committee on Toxicology, National Research Council. Recommendations for the Prevention of Lead Poisoning in Children. *Nutrition Reviews* 34(11): 321-327, 1977.

Cousins, Norman. *Anatomy of An Ailment.* New York: WW Norton & Co., 1978.

Crapper, DR, et al. *Science News* vol. 219, no. 5, Oct. 1, 1979.

Creason, John, et al. Trace Elements in Hair as Related to Exposure in Metropolitan New York. *Clinical Chemistry* 21(4):603-612, 1975.

Cunningham, AS. Lymphomas and Animal-Protein Consumption. *Lancet* 2:1184-1186, 1976.

Curran, JW. AIDS—Two Years Later. *New England Journal of Medicine* 309:609-611, 1983.

Daly, JC, et al. Iodine-129 Levels in Milk and Water Near a Nuclear Fuel Reprocessing Plant. *Health Physics* 26:333-342, 1974.

David, O, et al. The Relationship of Hyperactivity to Moderately Elevated Lead Levels. *Archives of Environmental Health* 38(6):335-359, 1983.

De Barreiro, OC. Effect of Cysteine on 5-Aminolaevulinate Hydrolase from Liver in Two Cases of Experimental Intoxication. *Biochemical Pharmacology* 18:2267, 1969.

Diet and Immunity. *American Journal of Clinical Nutrition* 32:1260-1266,1979. Donaldson, SS, et al. Radiation Enteritis in Children. *Cancer* 35:1167-1178, 1975.

Dufty, William. *Sugar Blues.* New York: Warner Books, 1975. Effect of Fluorine on Dental Caries. *Journal of the American Dental Association* 31:1360-1363, 1944.

Eggleston, David W. Effect of Dental Amalgam and Nickel Alloys on T-Lymphocytes: Preliminary Report. *Journal of Prosthetic Dentistry* 51(5):617-623.

Falck, F, et al. Protein Excretion Patterns in Cadmium-Exposed Individuals. *Archives of Environmental Health* 39(2):69-73, 1982.

Ferris, BJ. Health Effects of Exposure to Low Levels of Regulated Air Pollutants. *Air Pollution Control Association. Journal.* 28:422, 1978.

Freeman, Leslie J. *Nuclear Witnesses: Insiders Speak Out.* New York: WW Norton & Co., 1982.

Gabler, WL, and Leong, PA. Fluoride Inhibition of Polymorphonuclear Leukocytes. *Journal of Dental Research* 48(9):1933-1939, 1979.

G.A.O. Report Blasts Raw Meat and Poultry as Unsafe. *Food Engineering,* June 1979.

Gerber, G, and Altman, K. *Radiation Biochemistry.* New York: Academic Press, 1970.

Gilfillian, SC. Lead Poisoning and the Fall of Rome. *Journal of Occupational Medicine.* 7:53-55, 1965.

Golden, Michael, et al. The Effect of Zinc on Thymus of Recently Malnourished Children. *Lancet* 2:1057, 1977.

Goldwater, LJ. Mercury in the Environment. *Scientific American,* pp. 15-18, May 1971.

Grosch, DS, and Hopwood, LE. *Biological Effects of Radiation.* New York: Academic Press, 1979.

Gross, Robert, et al. Depressed Immunological Function in Zinc-Deprived Rats as Measured by Nitrogen Response of Spleen, Thymus and Peripheral Blood. *Arteriosclerosis,* May/June 1982.

Hall, R. Food Additives. *Nutrition Today,* p. 21, June/Aug. 1973.

Haveman, J. The Influence of pH on the Survival After X-Irradiation of Cultured Malignant Cells. *International Journal of Radiation Biology* 37:201-205, 1980.

High Fiber Diet: Taste is Hidden Attraction. *The New York Times,* p. C1, June 22, 1983.

Houton, L, and Sonnesso, G. Occupational Exposure and Cancer of the Liver. *Archives of Environmental Mental Health,* pp. 51-53, Jan./Feb. 1980.

Is Aluminum Harmless? *Nutrition Review* 38(7):242, 1980.

Iwao, S, et al. Variation of Cadmium Accumulation in Japanese. *Archives of Environmental Health* 38(3):156-161, 1983.

Jacobs, Leonard. Natural Ways to Survive a Meltdown. *East West Journal,* p. 62, June 1979.

Jacobson, Michael, and Brewster, L. *The Changing American Diet.* Washington: Center for Science in the Public Interest, 1978.

Jaffe, H, et al. Acquired Immune Deficiency Syndrome in the U.S.: The First 1000 Cases. *Journal of Infectious Diseases* 148(2):340-345, 1983.

Jensen, Bernard. *The Science and Practice of Iridology.* Orem, UT: Bi World Publishers, 1952.

Kaul, B, et al. Lead, Erythrocyte Protoporphyrin, and Ferritin Levels in Cord Blood. *Archives of Environmental Health* 38(5):296-299, 1983.

Koller, LD, et al. Antibody Suppression by Cadmium. *Archives of Environmental Health* 30(12):598-601, 1975.

Kushi, Aveline, and Esko, Wendy. *The Changing Seasons Macrobiotic Cookbook.* Garden City Park, NY: Avery Publishing Group, 1985.

Kushi, Michio. *The Book of Macrobiotics: The Universal Way of Health and Happiness.* Tokyo: Japan Publications, 1977.

Kushi, Michio. *Michio Kushi on the Greater View: Collected Thoughts and Ideas on Macrobiotics and Humanity.* Garden City Park, NY: Avery Publishing Group, 1985.

Kushi, Michio, and Jack, Alex. *The Cancer-Prevention Diet: Michio Kushi's Nutritional Blueprint for the Relief and Prevention of Disease.* New York: St. Martin's Press, 1983.

Lacourly, G. Relationship Between Radio-Cesium Contamination of Beef and Milk. *Health Physics* 21:793-802, 1971.

Landrigan, Philip. Summary of Work Group on Renal Disease. *Archives of Environmental Health* 39(3):252, 1984.

Landrigan, PJ, et al. Neuropsychologic Dysfunction in Children **with** Chronic Low-Level Lead Absorption. *Lancet,* pp. 708-712, 1975.

Lapp, Ralph. Nevada Test Fallout and Radioiodine in Milk. *Science,* pp. 756-758, Sept. 7, 1962.

Larsen, RB, and Oldham, RD. Plutonium in Drinking Water: Effects of Chlorination on Its Maximum Permissible Concentration. *Science,* 201:1008-1009, 1978.

Lead Persists as Threat to Young. *The New York Times,* p. Cl, May 13,1980.

Lenihan, JM, Ed. *Strontium Metabolism.* New York: Academic Press, 1967.

Lerner, J, and Musil, J. Cadmium Influence in the Excretion of Sodium by the Kidneys. *Experientia* 27:902, 1971.

Livingstone, Churchill. The Fate of Vitamins in Food Processing Operations. *Proceedings of the University of Nottingham Residential Seminar on Vitamins,* p. 73, 1971.

Lochiskar, M, et al. The Effect of Quantity and Quality of Dietary Fat on the Immune System. *Journal of Nutrition* 113:951, 1983.

Lower Lead Contamination Tied to Child Brain Damage. *The New York Times,* p. A18, March 29, 1979.

Mahaffey, Kathryn. Nutritional Factors in Lead Poisoning. *Nutrition Reviews* 39(10):353, 1981.

Mann, John. *Secrets of Life Extension.* New York: Bantam Books, 1980.

Marecek, J, et al. Low-Level Lead Exposure in Childhood Influence Neuro Psychological Performance. *Archives of Environmental Health* 38(6):355-359, Nov./Dec. 1983.

Martell, Edward A. *Cesium-137 From the Environment to Man: Metabolism and Dose.* NCRP Report #52. Washington: National Council on Radiation Protection and Measurements, 1977.

Marx, Jean. Aging Research: Pacemakers for Aging? *Science* 186:1196-1197, 1974.

Marx, Jean. Thymic Hormones: Inducers of T Cell Maturation. *Science* 187:1183, 1975.

Mayo, Anna. The Principle of Uncertainty. *New York Nat ive,* March 10,1986.

Mertz, W, et al. The Relation of Other Selected Trace Elements to Health and Disease. *Geochemistry and the Environment* 2:40-53, 1977.

Mineralab. *Mercury Toxicity.* Hayward, CA: Mineralab, Inc., 1976.

Mineralab. *Mineral Metabolism: A Report on Fourteen Minerals.* Hayward, CA: Mineralab, Inc., 1976.

Moreau, T, et al. Blood Cadmium Levels in a General Male Population with Special Reference to Smoking. *Archives of Environmental Health* 38(3):163-167, 1983.

National Research Council Committee on Lead in the Human Environment. *Lead in the Environment.* Washington: National Academy Press, 1980.

Needleman, H, et al. Deficits in Psychological and Classroom Performance of Children with Elevated Dentine Lead Levels. *New England Journal of Medicine* 300(13):689-695, 1979.

New Research on the Vegetarian Diet. *The New York Times,* p. Cl, Oct. 8, 1983.

Ohsawa, George. *Cancer and the Philosophy of the Far East.* Oroville, CA: George Ohsawa Macrobiotic Foundation, 1981.

Ophir, Orna, et al. Low Blood Pressure in Vegetarians: The Possible Role of Potassium. *American Journal of Clinical Nutrition* 37:755-762, 1983.

Passwater, Richard. *Supernutrition for Healthy Hearts.* New York: Dial Press, 1977.

Passwater, Richard, and Cranton, Elmer. *Trace Elements, Hair Analysis and Nutrition.* New Canaan, CT: Keats, 1983.

Perry, M, et al. Effect of a Second Metal on Cadmium Induced Hypertension. *Archives of Environmental Health* 38(2):80, 1983.

Perry, M, et al. Inhibition of Cadmium Induced Hypertension in Rats. *Science of the Total Environment* 14:153-166, 1980.

Pfeiffer, Carl C. *Mental and Elemental Nutrients.* New Canaan, CT: Keats, 1975.

Pfeiffer, Carl C. *Zinc and Other Micro-Nutrients.* New Canaan, CT: Keats, 1978.

Potera, Carol, et al. Vitamin B_6 Deficiency in Cancer Patients. *American Journal of Clinical Nutrition* 30:1677-1679, 1977.

Price, Weston A. *Nutrition and Physical Degeneration: A Comparison of Primitive and Modern Diets and Their Effects.* Santa Monica, CA: Price-Pottenger Nutrition Foundation, 1977.

Pritikin, Nathan. *The Pritikin Permanent Weight Loss Manual.* New York: Bantam Books, 1982.

Rasnoff, J. Congress Toughens Drinking Water Rules. *Science News* 129(22):341, 1986.

Residue of Chemicals in Meat Leads to Debate on Hazards. *The New York Times,* p. 1, March 15, 1983.

Rodale, Jl. *The Complete Book of Food and Nutrition.* Emmaus, PA: Rodale Press, 1966.

Rosner, Fred, and Giron, Jose. Immune Deficiency in Children (letter to the editor). *Journal of the American Medical Association* 250(22):3046, 1983.

Sacks, FM, Castelli, S, and Kass, EH. Plasma Lipids and Lipoproteins in Vegetarians and Controls. *New England Journal of Medicine* 292:1148-1151,1975.

Sacks, FM, Rosner, B, and Kass, EH. Blood Pressure in Vegetarians. *American Journal of Epidemiology* 100(5):390-398, 1974.

Schauss, Alexander. *Diet, Crime and Delinquency.* New York: Simon and Schuster, 1980.

Schroeder, HA, et al. *Trace Elements and Man.* Old Greenwich, CT: Devin-Adair, 1973.

Simpson, Robert E, et al. Projected Dose Commitment from Fallout Contamination in Milk Resulting from the 1976 Chinese Atmospheric Nuclear Weapons Test. *Health Physics* 40:741-743, May 1981.

Six, Kathryn M, and Goyer, Robert A. Experimental Enhancement of Lead Toxicity by Low Dietary Calcium. *Journal of Laboratory and Clinical Medicine* 76:933-942, 1970.

Souadjian, JV, et al. Morphologic Studies of the Thymus in Human Neoplasia. *Cancer* 23:619, 1969.

Souadjian, JV, et al. Thymoma and Cancer. *Cancer* 22:1221, 1968.

Spyker, Joan. Assessing the Impact of Low-Level Chemicals on Development: Behavioral and Latent Effects. *Federation of American Societies for Experimental Biology: Federation Proceedings* 34(9):1835-1844, 1975.

Spyker, Joan. Subtle Consequences of Methylmercury Exposure. *Teratology* 5:267, 1972.

Sternglass, Ernest J. The Implications of Chernobyl for Human Health. *International Journal of Biosocial Research* 8:7-36, 1986.

Sternglass, Ernest J. Personal communication, Jan. 1983.

Sternglass, Ernest J, and Scheer, J. Radiation Exposure of Bone Marrow Cells to Strontium-90 During Early Development as a Possible Cofactor in the Etiology of AIDS. Paper presented at the 1986 Annual Meeting of the American Association for the Advancement of Science (AAAS). Philadelphia: May 1986.

Sugimura, T, and Sato, S. Mutagens-Carcinogens in Foods. *Cancer Research* 43:2415, 1983.

Sulphur Amino Acids and the Calciuretic Effect of Dietary Protein. *Nutrition Reviews* 39:127, 1981.

Taub, Harold J. *Keeping Healthy in a Polluted World.* New York: Harper and Row, 1974.

Taylor, Alfred, and Taylor, Nell. Effect of Sodium Fluoride on Tumor Growth. *Proceedings of the Society for Experimental Biology and Medicine* 119:252-255, 1965.

Tilden, JH. *Toxemia,* rev. ed. New Canaan, CT: Keats, 1981.

Turner, M, et al. Methylmercury in Populations Eating Large Quantities of Marine Fish. *Archives of Environmental Health* 36(6):367-378, 1980.

U.S. Department of Health, Education and Welfare. *Increased Lead Absorption and Lead Poisoning in Young Children: A Statement by the Centers for Disease Control,* March 1975.

Vernon, Riley. Psychoneuroendocrine Influence on Immunocompetence and Neoplasia. *Science* 212:1100-1109, 1981.

Vitale, J, et al. Blood Lead—An Inadequate Measure of Occupational Exposure. *Journal of Occupational Medicine* 17(3):155-156, 1975.

Vitale, J, and Broitman, S. Lipids and Immune Function. *Cancer Research* 41:3706-3710, 1981.

Waldbott, George. *Fluoridation, The Great Dilemma.* Lawrence, Kansas: Coronado Press, 1978.

Waldbott, George. *Health Effects of Environmental Pollutants.* St. Louis: CV Mosby, 1973.

Ward, GM. The Cs-137 Content of Beef From Dairy and Feedlot Cattle. *Health Physics* 21:95-100, 1965.

Wasserman, Harvey, and Solomon, Norman. *Killing Our Own: The Disaster of America's Experience with Atomic Radiation.* New York: Dell Publishing Co., 1982.

White, DJ. Histochemical and Histological Effects of Lead on the Liver and *Kidney of the Dog. British Journal of Experimental Pathology* 58:101-112,1977.

Yiamouyiannis, John. *Fluoride: The Aging Factor.* Delaware, OH: Health Action Press, 1983.

Yokel, RA. Hair as an Indicator of Excessive Aluminum Exposure. *Clinical Chemistry* 28(4):662-665, 1982.

Yudkin, John. *Sweet and Dangerous.* New York: Bantam Books, 1973.

Zarkower, A. Alterations in Antibody Response Induced by Chronic Inhalation of SO_2 and Carbon. *Archives of Environmental Health* 25:45, 1972.

Ziff, Sam. *The Toxic Time Bomb: Silver Dental Fillings.* New York: Aurora Press, 1984.

Zinc and Immunocompetence. *Nutrition Reviews* 38:288-289, 1981.

Chapter Five

Abrahamson, E, and Pezet, AW. *Body, Mind and Sugar.* New York: Pyramid Books, 1971.

Akizuki, Tatsuichiro. *Nagasaki 1945.* London: Quartet Books, 1981.

Ames, Bruce N. Dietary Carcinogens and Anticarcinogens: Oxygen Radicals and Degenerative Diseases. *Science* 23:1256-1263, 1983.

Balabukha, Vera, Ed. *Chemical Protection Against Ionizing Radiation.* Elmsford, NY: Pergamon Press, 1963.

Carroll, K. Experimental Evidence of Dietary Factors and Hormone-Dependent Cancers. *Cancer Research* 35:3374-3383, 1975.

Cheraskin, E, and Ringsdorf, WM. *Psychodietetics.* Briarcliff Manor, NY: Stein and Day, 1974.

Chu, JY, et al. Studies in Calcium Metabolism: Effect of Low Calcium and Variable Protein Intake on Human Calcium Metabolism. *American Journal of Clinical Nutrition* 28:1028-1035, 1975.

Committee on Food Protection, Food and Nutrition Board, National Academy of Sciences. *Radionuclides in Foods,* pp. 22-56. Washington: National Academy Press, 1973.

Correa, P. Epidemiological Correlations Between Diet and Cancer Frequency. *Cancer Research* 41:3685-3690, 1981.

Davison, KL, Sell, JL and Rose, RJ. Dieldren Poisoning of Chickens During Severe Dietary Restriction. *Bulletin of Environmental Contamination and Toxicology* 5:493-501, 1970.

Dietary Goals for the United States, 2 ed. Prepared by the staff of the Select Committee on Nutrition and Human Needs, United States Senate. Washington: U.S. Government Printing Office, 1977.

Florida Division of Health. Radiological Surveillance Around the Florida Power Corporation's Crystal River Power Reactor Site, 1971. *Radiological Health Data and Reports,* pp. 709-713, Dec. 1972.

Grosch, DS, and Hopwood, LE. *Biological Effects of Radiation.* New York: Academic Press, 1979.

Haas, Robert. *Eat to Win: The Sports Nutrition Bible.* New York: New American Library, Signet Books, 1983.

Kushi, Michio. *The Book of Macrobiotics: The Universal Way of Health and Happiness.* Tokyo: Japan Publications, 1977.

Lenihan,], Ed. *Strontium Metabolism.* New York: Academic Press, 1967.

National Council on Radiation Protection and Measurements. *Management of Persons Accidentally Contaminated with Radionuclides.* NCRP Report #65. Washington: US Government Printing Office, 1980.

National Research Council. *Diet, Nutrition, and Cancer.* Washington: National Academy Press, 1982.

Passwater, Richard, and Cranton, Elmer. *Trace Elements, Hair Analysis and Nutrition.* New Canaan, CT: Keats, 1983.

Pearson, Durk, and Shaw, Sandy. *Life Extension.* New York: Warner Books, 1983.

Yudkin, John. *Sweet and Dangerous.* New York: Bantam Books, 1972.

Chapter Six

Adams, C, and Bonnel, J. Administration of Stable Iodide as a Means of Reducing Thyroid Irradiation Resulting from Inhalation of Radioactive Iodine. *Health Physics* 7:127, 1962.

Akizuki, Tatsuichiro. *Nagasaki 1945.* London: Quartet Books, 1981.

Ames, Bruce N. Dietary Carcinogens and Anticarcinogens: Oxygen Radicals and Degenerative Diseases. *Science* 23:1256-1263, 1983.

Anderson, K, et al. Nutrient Regulation of Chemical Metabolism in Humans. *Federation of American Societies for Experimental Biology: Federation Proceedings* 44(1):130-133, 1985.

Anderson, K, et al. Nutritional Influences on Chemical Biotransformations in Humans. *Nutrition Reviews* 40:161-171, 1982.

Bailey, Herbert. *E: The Essential Vitamin.* New York: Bantam Books, 1983.

Balabukha, Vera, Ed. *Chemical Protection Against Ionizing Radiation.* Elmsford, NY: Pergamon Press, 1963.

Berger, Stuart. *Dr. Berger's Immune Power Diet.* New York: Signet Books, 1985.

Bianchini, F, and Corbetta, F. *The Complete Book of Fruits and Vegetables.* New York: Crown Publishers, Inc., 1976.

Borek, C, et al. Conditions for Inhibiting and Enhancing Effects of the Protease Inhibitor on X-Ray Induced Neoplastic Transformation in Hamster and Mouse Cells. *National Academy of Sciences of the United States of America. Proceedings. Biological Sciences.* 76:1800-1803, 1979.

Burkitt, D, et al. Effect of Dietary Fiber on Stools and Transit Times and Its Role in the Causation of Disease. *Lancet* 2:1408-1412, 1972.

Burkitt, D, and Teowell, H. *Refined Carbohydrate Foods and Disease: Some Implications of Dietary Fiber.* London: Academic Press, 1975.

Burkitt, DP, Walker, AR, and Painter, NS. Dietary Fiber and Disease. *Journal of the American Medical Association* 229:1068-1074, 1974.

Calabrese, Edward James. *Nutrition and Environmental Health: The Influence of Nutritional Status on Pollutant Toxicity and Carcinogenicity.* New York: John Wiley & Sons, 1980.

Carlisle, Edith. A Relationship Between Silicon, Magnesium and Fluorine in Bone Formation in the Chick. *Federation of American Societies for Experimental Biology: Federation Proceedings,* pp. 1488-1493, 1962.

Carlisle, Edith. A Skeletal Alteration Associated with Silicon Deficiency. *Federation of American Societies for Experimental Biology: Federation Proceedings,* pp. 3999-4003, 1962.

Carr, T, et al. Reduction in the Absorption and Retention of Dietary Strontium in Man by Alginate. *International Journal of Radiation Biology* 14:225,1968.

Colditz, GA, et al. Increased Green and Yellow Vegetable Intake and Lowered Cancer Deaths in an Elderly Population. *American Journal of Clinical Nutrition* 41(1):32-36, 1985.

Copeland, Edmund. Mechanisms of Radioprotection—A Review. *Photochemistry and Photobiology* 128:839-844, 1978.

Cunningham, AS. Lymphomas and Animal-Protein Consumption. *Lancet* 2:1184-1186, 1976.

Cysteamine as a Protective Agent. *Radiation Research* 82:74, 1979.

Davis, Adelle. *Let's Get Well.* New York: Harcourt, Brace & World, 1965.

Diet and Cancer—Can Food Make a Difference? (editorial). *East West Journal,* p. 48, March 1983.

Ershoff, BH. Antitoxic Effects of Plant Fiber. *American Journal of Clinical Nutrition* 27:1395, 1974.

Esko, Edward and Esko, Wendy. *Macrobiotic Cooking for Everyone.* Tokyo: Japan Publications, 1980.

Galton, Lawrence. *The Truth About Fiber in Your Food.* New York: Crown Publishers, 1976.

Gofman, John W. *Radiation and Human Health: A Comprehensive Investigation of the Evidence Relating Low-Level Radiation to Cancer and Other Diseases.* San Francisco: Sierra Club Books, 1981.

Graham, S, et al. Diet in the Epidemiology of Cancer of the Colon and Rectum. *Journal of the National Cancer Institute* 61:709-714, 1978.

Graham, S, and Mettlin, C. Fiber and Other Constituents of Vegetables in Cancer Epidemiology. *Nutrition and Cancer: Etiology and Treatment,* pp. 189-225, 1981.

Grosch, DS, and Hopwood, LE. *Biological Effects of Radiation.* New York: Academic Press, 1979.

Haas, Robert. *Eat to Win: The Sports Nutrition Bible.* New York: New American Library, Signet Books, 1983.

Harman, Denham. Free Radical Theory of Aging—Nutritional Implications. *Age* 1(4):145-152, 1978.

Harvey, RF, et al. Effects of Increased Dietary Fiber on Intestinal Transit. *Lancet* 1:815, 1973.

Haveman, J. The Influence of pH on the Survival After X-Irradiation of Cultured Malignant Cells. *International Journal of Radiation Biology* 37:201-205, 1980.

Hennekens, CH, et al. Vitamin A and Risk of Cancer. *Journal of Nutrition Education* 14:135-136, 1982.

Henriksen, T, et al. Transfer of Radiation-Induced Unpaired Spins From Proteins to Sulfur Compounds. *Radiation Research* 18:163-176, 1963.

High Fiber Diet: Taste is Hidden Attraction. *The New York Times,* p. C1, June 22, 1983.

Hirayama, T. Relationship of Soybean-Paste Soup Intake to Gastric Cancer Risk. *Nutrition and Cancer* 3(4):223-233, 1982.

Iritani, N, and Nogi, S. Effects of Hijiki and Wakame on Cholesterol Turnover in the Rat. *Atherosclerosis* 15:87-92, 1972.

Jacobson, Michael, and Brewster, L. *The Changing American Diet.* Washington: Center for Science in the Public Interest, 1978.

Kennedy, A, and Little, J. Effects of Protease Inhibitors on Radiation Transformation in Vitro. *Cancer Research* 41:2103-2108, 1981.

Kennedy, A, and Little, J. Protease Inhibitors Suppress Radiation Induced Malignant Transformation in Vitro. *Nature* 276:825-826, 1978.

King, M, and McCay, P. Modulation of Tumor Incidence and Possible Mechanisms of Inhibition of Mammary Carcinogenesis by Dietary Antiox-idants. *Cancer Research* 43:2485-2490, 1983.

Knuiman, JT, and West, CE. The Concentration of Cholesterol in Serum and in the Various Serum Lipoproteins in Macrobiotic, Vegetarian, and Non-Vegetarian Men and Boys. *Arteriosclerosis* 43:71-82, 1982.

Kozhokaru, AF, et al. Modification of Radiation Damage by Diabasic Sulfur-Containing Acids of a Phenol Series. *Radiobiologica* 22(4):545-548, 1982.

Kromhout, et al. Dietary Fiber and 10 Year Mortality From Coronary Heart Disease, Cancer and All Causes. *Lancet* 2, 518-521, Sept. 4, 1982.

Kushi, Michio. *The Book of Macrobiotics: The Universal Way of Health and Happiness.* Tokyo: Japan Publications, 1977.

Kvale, G, et al. Dietary Habits and Lung Cancer Risk. *International Journal of Cancer* 31:397-405, 1983.

Lappe, Frances Moore. *Diet for a Small Planet.* New York: Ballantine, 1971.

Lenihan, JM, Ed. *Strontium Metabolism.* New York: Academic Press, 1967.

Leonard, Thorn. Jesting Sea Vegetables. *East West Journal,* p. 16, Sept. 1983.

Liener, I. Significance for Humans of Biologically Active Factors in Soybeans and Other Food Legumes. *Journal of American Oil Chemists Society* 56:121—129, 1979.

Loub, WD, Wattenberg, LW, and Davis, DW. Aryl Hydrocarbon Hy-droxylase Induction in Rat Tissues by Naturally Occurring Indoles of Cruciferous Plants. *Journal of the National Cancer Institute* 54:985-988, 1975.

Lourou, M, and Lartigue, O. *Experientia* 6:25, 1950.

Lust, John, *The Herb Book.* New York: Benedict Lust Publications, 1974.

Mangelsdorf, Paul C. Wheat. *Scientific American,* July 1953.

Mann, John. *Secrets of Life Extension.* New York: Bantam Books, 1980.

Mechanism of Action of Aminothiol Protectors. *Nature* 213:363, 1967.

Moyer, Anne. *The Fiber Factor.* Emmaus, PA: Rodale Press, 1976.

Nagasawa, H, and Little, J. Effect of Tumor Promoters, Protease Inhibitors and Repair Processes on X-Ray Induced Sister Chromatid Exchanges in Mouse Cells. *National Academy of Sciences of the United States of America. Proceedings. Biological Sciences.* 76:1943-1947, 1979.

Nagata, C, and Yamaguchi, T. Electronic Structure of Sulfur Compounds and Their Protecting Action Against Ionizing Radiation. *Radiation Research* 73:430-439, 1978.

National Council on Radiation Protection and Measurements. *Management of Persons Accidentally Contaminated with Radionuclides.* NCRP Report #65. Washington: US Government Printing Office, 1980.

National Research Council. *Diet, Nutrition, and Cancer.* Washington: National Academy Press, 1982.

Newell, GR, and Ellison, NM, Eds. *Nutrition and Cancer: Etiology and Treatment* (Progress in Cancer Research and Therapy, vol. 17). New York: Raven Press, 1981.

Ohlson, MA, et al. Perspectives in Nutrition: Changes in Retail Market Food Supplies in the US in the Last Seventy Years in Relation to the Incidence of Coronary Heart Disease with Special Reference to Dietary Carbohydrates and Essential Fatty Acids. *American journal of Clinical Nutrition,* 14:169-178, 1964.

Palladino, M, et al. Irradiation-Induced Mortality: Protective Effect of Protease Inhibitors in Chickens and Mice. *International journal of Radiation Biology* 41:183-191, 1982.

Pantuck, EJ, et al. Effect of Brussels Sprouts and Cabbage on Drug Conjugation in Humans. *Clinical Pharmacology and Therapeutics* 35:161-169, 1984.

Pantuck, EJ, et al. Stimulatory Effect of Brussels Sprouts and Cabbage on Drug Metabolism. *Clinical Pharmacology and Therapeutics* 25:88, 1979.

Passwater, Richard, and Cranton, Elmer. *Trace Elements, Hair Analysis and Nutrition.* New Canaan, CT: Keats, 1983.

Pearson, Durk, and Shaw, Sandy. *Life Extension.* New York: Warner Books, 1983.

Petkau, A, and Chelack, WS. Radioprotective Effects of Cysteine. *International journal of Radiobiology* 25:321, 1974.

Peto, R, et al. Can Dietary Beta-Carotene Materially Reduce Human Cancer Rates? *Nature* 290:201-208, 1981.

Pfeiffer, Carl C. *Mental and Elemental Nutrients.* New Canaan, CT: Keats, 1975.

Pritikin, Nathan. *The Pritikin Permanent Weight Loss Manual.* New York: Bantam Books, 1982.

Pryor, William. Free Radicals in Biological Systems. *Scientific American,* pp. 70-83, August 1970.

Raloff, J. Tracing Disease to Trace Minerals. *Science News* 23:358, 1985.

Ray, RM, and Strasberg, SM. Origin, Chemistry, Physiological Effects and Clinical Importance of Dietary Fiber. *Clinical Investigative Medicine* 1:9-24,1978.

Reuben, David. *The Save Your Life Diet.* New York: Random House, 1975.

Sacks, FM, Castelli, S, and Kass, EH. Plasma Lipids and Lipoproteins in Vegetarians and Controls. New *England Journal of Medicine* 292:1148-1151,1975.

Sanner, T. Transfer of Radiation Energy From Macromolecules to Sulfur Compounds. *Radiation Research* 44:594-604, 1970.

Schoeters, G, et al. Influence of Ra-226 on Bone Marrow Stem Cells in Mice: Effect of Radium Decorporation by a Long Term Treatment with Sodium Alginate on Stem-Cell Damage. *Radiation Research* 82:74, 1979.

Shurtleff, William, and Aoyagi, Akiko. *The Book of Miso.* Brookline, MA: Autumn Press, 1976.

Simpson, HC, Mann, JI. Effect of High-Fiber Diet on Haemostatic Variables in Diabetes. *British Journal of Medicine* 1284:1609, 1982.

Skoryna, Stanley C. Reduction of Strontium Absorption in Man by the Addition of Alginate Derivative. *Nature* 216:1005, 1967.

Skoryna, Stanley C, et al. Studies on Inhibition of Intestinal Absorption of Radioactive Strontium. *Canadian Medical Association Journal* 91:285-288,1964.

Southgate, DA, and Bailey, E. A Guide to Calculating Intakes of Dietary Fiber. *Journal of Human Nutrition* 30:303-313, 1976.

Spector, H, and Calloway, D. *Proceedings of the Society for Experimental Biology and Medicine* 100:405, 1959.

Sporn, MB, and Newton, DL. Chemoprevention of Cancer with Retinoids. *Federation of American Societies for Experimental Biology: Federation Proceedings* 38:2528-2534, 1979.

Stantchev, et al. Administration of Granular Pectin to Workers Exposed to Lead. *Zeit. Ges Hygiene Gregebiet* 25:585-587, 1979.

Sugahara, T, et al. Studies on a Sulfhydrl Radioprotector of Low Toxicity. *Experientia* 27:53-61, 1977.

Suppression of Intestinal Absorption of Radiostrontium by Naturally Occurring Non-Absorbable Polyelectrolytes. *Nature* p. 205, March 13, 1965.

Sutton, A. A Reduction of Strontium Absorption by the Addition of Alginate to the Diet. *Nature* 216:1005, 1967.

Suzuki, YI, etal. Antitumor Effect of Seaweed. *Chemotherapy 28(2):165-170,* 1980.

Tannenbaum, A, and Silverstone, H. Effects of Varying the Portion of Protein in the Diet. *Cancer Research* 9:162-173, 1949.

Teas, J. The Consumption of Seaweed as a Protective Factor in the Etiology of Breast Cancer. *Medical Hypotheses* 7:601-613, 1981.

Tewfik, HH, et al. The Influence of Ascorbic Acid on Survival of Mice Following Whole Body X-Irradiation. *International Journal of Vitamin and Nutrition Research* 23:265-276, 1982.

Tkac, Debora. Fiber—How Much Do You Really Need? *Prevention,* pp. 113-127, Feb. 1983.

Triffit, JT. Binding of Calcium and Strontium by Alginates. *Nature* 217:457-458, 1968.

Troll, Walter. Blocking of Tumor Promotion by Protease Inhibitors, pp. 549-555 in *Cancer: Achievement, Challenges, and Prospects for the 1980s,* vol 1. Orlando, FL: Grune and Stratton, 1981.

Troll, Walter, et al. Inhibition of Carcinogenesis by Feeding Diets Containing Soybeans. *Proceedings of the American Association for Cancer Research* 20:265, 1979.

Vanderborght, OLJ, et al. Effect of Combined Alginate Treatments on the Distribution and Excretion of an Old Radiostrontium Contamination. *Health Physics* 35:255-258, 1978.

Waldron-Edward, D, Paul, TM, and Skoryna, S. Inhibition of Absorption of Radioactive Strontium by Alginic Acid Derivatives, in

Lenihan, JM, Ed., *Strontium Metabolism.* New York: Academic Press, 1967.

Wapnir, R, et al. Reduction of Lead Toxicity in the Kidney and the Small Intestine by Kaolin and Pectin in the Diet. *American Journal of Clinical Nutrition* 33:2303-2310, 1980.

Wattenberg, Lee W, et al. Dietary Constituents Altering the Responses to Chemical Carcinogens. *Federation of American Societies for Experimental Biology: Federation Proceedings* 35:1327-1331, 1976.

Wattenberg, Lee W, et al. Inhibition of Neoplasia by Minor Dietary Constituents. *Cancer Research* 43:2448-2453, 1983.

Wattenberg, Lee W, et al. Inhibition of Polycyclic Hydro Carbon Induced Neoplasia by Naturally Occurring Indoles. *Cancer Research* 38:1410-1413,1978.

Winn, DM, et al. Diet in the Etiology of Oral and Pharyngeal Cancer Among Women from the Southern US. *Cancer Research* 44:1216-1222, 1984.

Wood, L, et al. *Your Thyroid: A Home Reference.* Boston: Houghton Mifflin, 1982.

Wooley, DW. *The Biochemical Basis of Psychosis.* New York: John Wiley & Sons, 1962.

Wynder, EL, and Bross, ID. A Study of Etiological Factors: Cancer of the Esophagus. *Cancer* 14:389-413, 1961.

Yavelow, J, et al. Bowman-Birk Soybean Protease Inhibitor as an Anticar-cinogen. *Cancer Research* 43:2454-2459, 1983.

Chapter Seven

Airola, Paavo. *The Miracle of Garlic.* Phoenix, AZ: Health Plus Publishers, 1978.

Aly, HE. Oral Contraceptives and Vitamin B_6 Metabolism. *American Journal of Clinical Nutrition* 24:297-303, 1971.

Ames, Bruce N. Dietary Carcinogens and Anticarcinogens: Oxygen Radicals and Degenerative Diseases. *Science* 23:1256-1263, 1983.

Anderson, K, et al. The Effects of Increasing Weekly Doses of Ascorbate on Certain Cellular and Humoral Immune Functions. *American Journal of Clinical Nutrition* 33:71-76, 1980.

Ascorbic Acid: Its Ability to Induce Immunity Against Some Cancers in Mice. *Physiological Chemistry and Physics* 13:325, 1981.

Bailey, Herbert. £: *The Essential Vitamin.* New York: Bantam Books, 1983.

Bauerstock, KF. Radioprotection by Vitamin C. *British Journal of Radiology* 52:592-593, 1979.

Bingham, S, et al. Dietary Fibre and Regional Large-Bowel Cancer Mortality in Britain. *British Journal of Cancer* 40:456-463, 1979.

Bjelke, E. Dietary Vitamin A and Human Lung Cancer. *International Journal of Cancer* 15:561-565, 1975.

Braverman, ER, and Pfeiffer, CC. Essential Trace Elements and Cancer. *The Journal of Orthomolecular Psychiatry* 11(1):28-41, 1982.

Burkitt, DP, and Trowell, HC. *Refined Carbohydrate Foods and Disease: Some Implications of Dietary Fiber.* London: Academic Press, 1975.

Calabrese, Edward James. *Nutrition and Environmental Health: The Influence of Nutritional Status on Pollutant Toxicity and Carcinogenicity.* New York: John Wiley & Sons, 1980.

Cameron, E, and Pauling, L. Supplemental Ascorbate in the Supportive Treatment of Cancer: Prolongation of Survival Times in Terminal Human Cancer. *National Academy of Sciences of the United States of America. Proceedings-Biological Sciences.* 73(10):3685-3689, 1976.

Chandra, RK. Immunodeficiency in Under nutrition and Over nutrition. *Nutrition Reviews* 39(6):225-231, 1981.

Chandra, RK. Serum Thymic Hormone Activity in Protein Energy Malnutrition. *Clinical and Experimental Immunology* 38:228-230, 1979.

Combridge, CD. The Effect of Pyridoxine Deficiency on Certain Organs of the Rat. *British Journal of Nutrition* 10:347, 1956.

Cook, MG, and McNamara, P. Effect of Dietary Vitamin E on Dimethylhy-drazine-Induced Colonic Tumors in Mice. *Cancer Research* 40:1329, 1980.

Copeland, Edmund. Mechanisms of Radioprotection—A Review. *Photochemistry and Photobiology* 128:839-844, 1978.

Curtis, HJ. Radiation and Aging. *Society for Experimental Biology and Medicine. Proceedings.* Symposium 21:51, 1967.

Dardenne, M, et al. Contribution of Zinc and Other Metals to the Biological Activity of the Serum Thymic Factor. *National Academy of Sciences of the United States of America. Proceedings. Biological Sciences.* 79:5370-5373, 1982.

Dardenne, M, et al. In Vivo and In Vitro Studies of Thymolin in Marginally Zinc-Deficient Mice. *European Journal of Immunology* 14:454-458, 1984.

Delves, H, et al. Copper and Zinc Concentration in the Plasma of Leukemic Children. *British Journal of Haematology* 24:525-531, 1973.

Depasquale-Jardieu, P, and Fraker, PJ. The Role of Corticosterone in the Loss of Immune Function in the Zinc-Deficient A/J Mouse. *Journal of Nutrition* 109:1847-1855, 1979.

Diplock, AT. Possible Stabilizing Effect of Vitamin E on Microsomal Membrane Bound, Selenide-Containing Proteins and Drug Metalizing Enzyme Systems. *American Journal of Clinical Nutrition* 27:995-1004, 1974.

Dolar, SG, and Keeney, DR. Availability of Cu, Zn and Mn in Soils. *Journal of the Science of Food and Agriculture* 22:273-286, 1972.

Gailani, SD, et al. Clinical and Biochemical Studies of Pyridoxine Deficiency in Patients with Neoplastic Diseases. *Cancer* 21:975, 1968.

Golden, Michael, et al. Effect of Zinc on Thymus of Recently Malnourished Children. *Lancet* 2:1057, Nov. 1977.

Graham, S, and Mettlin, C. Fiber and Other Constituents of Vegetables in Cancer Epidemiology. *Nutrition and Cancer: Etiology and Treatment,* pp. 189-225, 1981.

Gregory, NL. The Mechanism of Radioprotection by Vitamin C. *British Journal of Radiology* 51:473-474.

Gross, R. Depressed Immunological Function in Zinc Deprived Rats as Measured by Mitogen Response of Spleen, Thymus and Peripheral Blood. *American Journal of Clinical Nutrition* 32:1260-1265, 1979.

Gruberg, E, and Raymond, S. *Beyond Cholesterol: Vitamin B_6, Atherosclerosis, and Your Heart.* New York: St. Martin's Press, 1981.

Haenszel, W, and Correa, P. Developments in the Epidemiology of Stomach Cancer Over the Past Decade. *Cancer Research* 35:3452-3459, 1975.

Harman, Denham. Free Radical Theory of Aging: Nutritional Implications. *Age* 1:145-152, 1978.

Harman, Denham. Prolongation of Lifespan and Inhibition of Spontaneous Cancer by Antioxidants. *Journal of Gerontology* 16:247, 1981.

Hirayama, T. Relationship of Soybean-Paste Soup Intake to Gastric Cancer Risk. *Nutrition and Cancer* 3(4):223-233, 1982.

Horwitt, MK. Vitamin E: A Reexamination. *American Journal of Clinical Nutrition* 29:569, 1976.

Is Dietary Carotene an Anti-Cancer Agent? *Nutrition Reviews* 40(9):257-261, 1982.

Kasuya, M. The Effect of Vitamin E on the Toxicity of Alkyl Mercurials on Nerve Tissue in Culture. *Toxicology and Applied Pharmacology* 32:347, 1975.

Kratzer, FH, and Williams, DE. The Relation of Pyridoxine to the Growth of Chicks Fed Rations Containing Linseed Oil. *Journal of Nutrition* 36:297,1948.

Lucy, JE. Functional and Structural Aspects of Biological Role For Vitamin E in the Control of Membrane Permeability and Stability. *Annals of the New York Academy of Sciences,* pp. 203-204, 1972.

Luhby, AL, et al. Pyridoxine and Oral Contraceptives. *Lancet* 2:1083,1970.

MacLennan, R., et al. Diet, Transit Time, Stool Weight, and Colon Cancer in Two Scandinavian Populations. *American Journal of Clinical Nutrition* 31:S239-S242, 1978.

Marx, J. Estrogen Drugs: Do They Increase the Risk of Cancer? *Research News Science* 191:838, 1976.

Miller, LT, et al. B_6 Metabolism in Women Using Oral Contraceptives. *American Journal of Clinical Nutrition* 27:797-805, 1974.

O'Connor, MK, and Malone, JF. A Radioprotective Effect of C Observed in Chinese Hamster Ovary Cells. *British Journal of Radiology* 50:587-591,1977.

Pearson, Durk, and Shaw, Sandy. *Life Extension.* New York: Warner Books, 1982.

Pekarek, RS, et al. Abnormal Cellular Immune Responses During Acquired Zinc Deficiency. *American Journal of Clinical Nutrition* 32:1466-1471, 1979.

Peto, R, et al. Can Dietary Beta-Carotene Materially Reduce Human Cancer Rates? *Nature* 290:201-206, 1981.

Pfeiffer, Carl C. *Mental and Elemental Nutrients.* New Canaan, CT: Keats, 1975.

Pfeiffer, Carl C. *Zinc and Other Micro-Nutrients.* New Canaan, CT: Keats, 1978.

Roehm, JN, et al. The Influence of Vitamin E on the Lung Fatty Acids of Rats Exposed to Ozone. *Archives of Environmental Health* 24:237, 1972.

Rosenberg, SJ, and Bennett, JM. Pyridoxine Responsive Anemia. *New York State Journal of Medicine* 96:1430-1433, 1969.

Sandstead, H. Zinc Nutrition in the U.S. *American Journal of Clinical Nutrition* 26:1251, 1973.

Sauberlich, HE, et al. Biochemical Assessment of the Nutritional Status of B$_6$ in the Human. *American Journal of Clinical Nutrition* 25:629-642, 1972.

Schroeder, HA. Losses of Vitamins and Trace Minerals Resulting From Processing and Preservation of Foods. *American Journal of Clinical Nutrition* 24:562-573, 1971.

Siegel, B, and Morton, JI. Vitamin C and the Immune Response. *Experientia* 33:393, 1977.

Sporn, MB, and Newton, DL. Chemoprevention of Cancer with Retinoids. *Federation of American Societies for Experimental Biology: Federation Proceedings* 38:2528-2534, 1979.

Sullivan, JF, and Lamkford, HG. Urinary Excretion of Zinc in Alcoholism and Post Alcoholic Cirrhosis. *American Journal of Clinical Nutrition* 10:153-157, 1962.

Suzuki, Y, et al. Antitumor Effect of Seaweed. *Chemotherapy 28(2):165-170,* 1980.

Taper, J, et al. Effects of Zinc Intake on Copper Balance in Adult Females. *American Journal of Clinical Nutrition* 33:1077-1082, 1980.

Tappel, AL. Vitamin E Spares the Parts of the Cell and Tissues From Free Radical Damage. *Nutrition Today* 8:4, 1973.

Teas, J. The Consumption of Seaweed as a Protective Factor in the Etiology of Breast Cancer. *Medical Hypotheses* 7:601-613, 1981.

Triffit, JT. Binding of Calcium and Strontium by Alginates. *Nature* 217:457-458, 1968.

Troll, Walter. Blocking of Tumor Promotion by Protease Inhibitors, pp. 549-555 in *Cancer: Achievements, Challenges and Prospects for the 1980s,* vol. 1. Orlando, FL: Grune and Stratton, 1981.

Wade, S, et al. Thymulin (Zn Facteur Thymique Serique) Activity in Anorexia Patients. *American journal of Clinical Nutrition* 42:275-280, 1985.

Wattenberg, Lee W. Inhibition of Carcinogenic and Toxic Effects of Polycyc-lic Hydrocarbons by Phenalic Antioxidants and Ethoxyquin. *Journal of the National Cancer Institute* 48:1425-1430, 1972.

Weisburger, JH, and Raineri, R. Dietary Factors and the Etiology of Gastric Cancer. *Cancer Research* 35:3469-3474, 1975.

Zinc and Immunocompetence. *Nutrition Reviews* 38:288-289, 1981.

Chapter Eight

Kushi, Aveline, and Esko, Wendy. *The Changing Seasons Macrobiotic Cookbook.* Garden City Park, NY: Avery Publishing Group, 1985.

Kushi, Michio, with Blauer, Stephen. *The Macrobiotic Way: The Complete Macrobiotic Diet and Exercise Book.* Garden City Park, NY: Avery Publishing Group, 1985.

Chapter Nine

Institute for Defense and Disarmament Studies. *Peace Resource Book: A Comprehensive Guide to Issues, Groups, and Literature.* Cambridge, MA: Ballinger, 1986.

Jackson, Thomas P. Optimizing the Performance of Direct-Reading Dosimeters. *Health Physics* 49(1):49-54, 1985.

Nuclear Issues and Radiation Effects

Ackland, Len, and McGuire, Steven. Assessing *the Nuclear Age.* Chicago: Bulletin of the Atomic Scientists, 1986.
A guidebook to the nuclear era, with contributions by over forty scientists and historians.
Arkin, William M and Fieldhouse, Richard W. *Nuclear Battlefields: Global Links in the Arms Race.* Cambridge, MA: Ballinger, 1985.
All the details on nuclear weapons.
Bartlett, Donald L and Steele, James B. *Forevermore: Nuclear Waste in America.* New York: WW Norton & Co., 1985.
Berger, John J. *Nuclear Power: The Unviable Option.* Palo Alto, CA: Ramparts Press, 1976.
Brodeur, Paul. *The Zapping of America: Microwaves, Their Deadly Risk, and the Cover-Up.* New York: WW Norton & Co., 1977.
Browne, Corinne, and Munroe, Robert. *Time Bomb: Understanding the Threat of Nuclear Power.* New York: William Morrow & Co., 1981.
Bruland, W, et al. *Radioecological Assessment of the Wyhl Nuclear Power Plant* (the "Heidelberg Report"). Heidelberg, West Germany: Department of Environmental Protection of the University of Heidelberg, rev. 1979.
This important report by a team of fourteen West German scientists from Heidelberg University shows that the potential radiation dose to humans from nuclear power plants is vastly underestimated. The Heidelberg Report is available (be sure to request an English translation) from:

IFEU—Institut fur Energie und Umweltforschung
Heidelberg e.V.
Im Sand 5
D-6900 Heidelberg
West Germany

Caldicott, Helen. *Missile Envy,* rev. ed. New York: Bantam Books, 1986. *Must* reading.

Caldicott, Helen. *Nuclear Madness: What You Can Do!* New York: Bantam Books, 1979.

Croall, Stephen, and Kaianders. *The Anti-Nuclear Handbook.* New York: Random House, Pantheon Books, 1978.

Presents information concerning the risks of nuclear technology in comic-book style.

Faulkner, Peter, Ed. *The Silent Bomb: A Guide to the Nuclear Energy Controversy.* New York: Random House, 1977.

Freeman, Leslie J. *Nuclear Witnesses: Insiders Speak Out.* New York: WW Norton & Co., 1982.

Contains sixteen interviews with people related to the nuclear industry, from a pipefitter in a power plant to scientists Ernest Sternglass and John Gofman. *Nuclear Witnesses* also contains an excellent chronology of events in the history of nuclear power. With an Afterword by Helen Caldicott.

Fuller, John G.*The Day We Bombed Utah.* New York: NAL Books, 1984.

The Conqueror, starring John Wayne, Dick Powell, Susan Powell, and Agnes Moorehead was filmed in Utah in 1954. All four stars—and nearly half the cast—contracted cancer. The stars and half the others who got cancer died from it. This book tells the tale.

Gofman, John W. *Radiation and Human Health: A Comprehensive Investigation of the Evidence Relating Low-Level Radiation to Cancer and Other Diseases.* San Francisco: Sierra Club Books, 1981.

An analysis of radiation's effects on humans by one of the world's leading authorities.

Gofman, John W, and O'Connor, Egan. *X-Rays: Health Effects of Common Exams.* San Francisco: Sierra Club Books, 1985.

Gofman, John W, and Tamplin, Arthur R. *Poisoned Power: The Case Against Nuclear Power Plants Before and After Three Mile Island.* Emmaus, PA: Rodale Press, 1979.

Information on nuclear power and radiation simply explained.

Grossman, Karl. *Cover-Up: What You Are Not Supposed to Know About Nuclear Power.* Sag Harbor, NY: Permanent Press, 1982.

A well-documented book presenting evidence that nuclear power poses great dangers.

Grossman, Karl. *Power Crazy.* New York: Grove Press, 1986.

The story of LILCO (the Long Island Lighting Company) and the Shoreham, NY nuclear plant and a study of the nuclear establishment.

Gyorgy, Anna, and Friends. No *Nukes: Everyone's Guide to Nuclear Power.* Boston: South End Press, 1979.

Hilgartner, Stephen, Bell, Richard C, and O'Connor, Rory. *Nukespeak: The Selling of Nuclear Technology in America.* New York: Viking Penguin Books, 1982.

A description of the language of euphemism and the public relations campaign used to sell nuclear technology.

Kennan, George F. *The Nuclear Delusion: Soviet-American Relations in the Atomic Age.* New York: Random House, Pantheon Books, 1982.

Lipshutz, Ronnie D. *Radioactive Waste: Politics, Technology, and Risk. A* Report of the Union of Concerned Scientists. Cambridge, MA: Ballinger, 1980.

Loeb, Paul. *Nuclear Culture.* Philadelphia: New Society Publishers, 1986.

An account of the growth of the Hanford, Washington nuclear plant and nuclear culture.

Miller, Richard L. *Under the Cloud: The Decades of Nuclear Testing.* New York: Macmillan, The Free Press, 1986.

A detailed report on America's atmospheric bomb tests of the 1950s and 1960s, including maps that trace the path of clouds and radioactive fallout.

Nader, Ralph, and Abbots, John. *The Menace of Atomic Energy.* New York: WW Norton & Co., 1977.

Pauling, Linus. *No More War.* New York: Dodd, Mead & Co., 1958.

The Peace Resource Book: A Comprehensive Guide to Issues, Groups and Literature, compiled by the Institute for Defense and Disarmament Studies (I.D.D.S.). Cambridge, MA: Ballinger, 1986.

Schell, Jonathan. *The Fate of the Earth. New* York: Alfred A. Knopf, 1982.

Shutdown: Nuclear Power on Trial. Transcript of proceedings in Federal Court. Summertown, TN: The Book Publishing Co., 1979.

The testimony of hearings in Federal court, October 2,1978. Ernest Sternglass and John Gorman are cross-examined.

Sternglass, Ernest J. *Secret Fallout: Low-Level Radiation from Hiroshima to Three Mile Island.* New York: McGraw-Hill, 1981.

The book represents twenty years of work to educate the public concerning the devastating health effects of low-level radiation.

Szilard, Leo. *The Voice of the Dolphins.* New York: Simon and Schuster, 1961.

Wasserman, Harvey, and Solomon, Norman. *Killing Our Own: The Disaster of America's Experience with Atomic Radiation.* New York: Dell Publishing Co., 1982.

Must reading. Documents the national health disaster resulting from nuclear power. The result of years of investigative research. Extensively documented.

Wick, O.J., Ed. *Plutonium Handbook: A Guide to the Technology.* La Grange Park, IL: American Nuclear Society, 1980.

Contains twenty-three chapters on all aspects of plutonium.

Williams, Robert C, and Cantelon, Philip L. *The American Atom: A Documentary History of Nuclear Politics from the Discovery of Fission to the Present (1939-1984).* Philadelphia: University of Pennsylvania Press, 1984.

A collection of primary-source documents pertaining to nuclear history.

Periodic Chart of the Elements

THE SYMBOL Shown in the middle of each block directly below the name of the element.

THE ATOMIC WEIGHT Directly below the symbol for each element the atomic weight is shown. The values are taken from the official Report on Atomic Weights Cr. J. Amer. Chem. Soc. 84, 4193 (1976). For elements not listed in the Report the mass number of the longest lived isotope is shown in brackets.

THE ATOMIC NUMBER Shown in the upper left hand corner.

ELECTRONIC CONFIGURATION Shown at the upper right as a group of numerals. When read downward they indicate the number of electrons normally found in successive energy levels.

NOBLE GASES

| IA | IIA | IIIA | IVA | VA | VIA | VIIA | VIIIA | | | IB | IIB | IIIB | IVB | VB | VIB | VIIB |

Period 1:
1 H 1.0079 — 2 He 4.00260

Period 2:
3 Li 6.941 — 4 Be 9.01218 — 5 B 10.81 — 6 C 12.011 — 7 N 14.0067 — 8 O 15.9994 — 9 F 18.99840 — 10 Ne 20.179

Period 3:
11 Na 22.98977 — 12 Mg 24.305 — 13 Al 26.98154 — 14 Si 28.0855 — 15 P 30.97376 — 16 S 32.06 — 17 Cl 35.453 — 18 Ar 39.948

Period 4:
19 K 39.0983 — 20 Ca 40.08 — 21 Sc 44.9559 — 22 Ti 47.90 — 23 V 50.9415 — 24 Cr 51.996 — 25 Mn 54.9380 — 26 Fe 55.847 — 27 Co 58.9332 — 28 Ni 58.70 — 29 Cu 63.546 — 30 Zn 65.38 — 31 Ga 69.72 — 32 Ge 72.59 — 33 As 74.9216 — 34 Se 78.96 — 35 Br 79.904 — 36 Kr 83.80

Period 5:
37 Rb 85.4678 — 38 Sr 87.62 — 39 Y 88.9059 — 40 Zr 91.22 — 41 Nb 92.9064 — 42 Mo 95.94 — 43 Tc (97) — 44 Ru 101.07 — 45 Rh 102.9055 — 46 Pd 106.4 — 47 Ag 107.868 — 48 Cd 112.41 — 49 In 114.82 — 50 Sn 118.69 — 51 Sb 121.75 — 52 Te 127.60 — 53 I 126.9045 — 54 Xe 131.30

Period 6:
55 Cs 132.9054 — 56 Ba 137.33 — 57 *La 138.9055 — 72 Hf 178.49 — 73 Ta 180.9479 — 74 W 183.85 — 75 Re 186.207 — 76 Os 190.2 — 77 Ir 192.22 — 78 Pt 195.09 — 79 Au 196.9665 — 80 Hg 200.59 — 81 Tl 204.37 — 82 Pb 207.2 — 83 Bi 208.9804 — 84 Po (209) — 85 At (210) — 86 Rn (222)

Period 7:
87 Fr (223) — 88 Ra 226.0254 — 89 **Ac (227) — 104 Unq (261) — 105 Unp (262) — 106 Unh (263) — 107 Uns (262)

* Lanthanide Series:
58 Ce 140.12 — 59 Pr 140.9077 — 60 Nd 144.24 — 61 Pm (147) — 62 Sm 150.4 — 63 Eu 151.96 — 64 Gd 157.25 — 65 Tb 158.9254 — 66 Dy 162.50 — 67 Ho 164.9304 — 68 Er 167.26 — 69 Tm 168.9342 — 70 Yb 173.04 — 71 Lu 174.97

** Actinide Series:
90 Th 232.0381 — 91 Pa 231.0359 — 92 U 238.029 — 93 Np 237.0482 — 94 Pu (244) — 95 Am (243) — 96 Cm (247) — 97 Bk (247) — 98 Cf (251) — 99 Es (254) — 100 Fm (257) — 101 Md (258) — 102 No (259) — 103 Lr (260)

The Chemical Elements and Their Symbols

The following listing includes all of the known chemical elements and their standard abbreviations (symbols). The names are listed alphabetically for ease of reference.

Element	Symbol	Element	Symbol
actinium	Ac	fluorine	F
alumninum	Al	francium	Fr
americium	Am	gadolinium	Gd
antimony	Sb	gallium	Ga
argon	Ar	germanium	Ge
arsenic	As	gold	Au
astatine	At	hafnium	III
barium	Ba	helium	He
berkelium	Bk	holmium	Ho
beryllium	Be	hydrogen	H
bismuth	Bi	indium	In
boron	B	iodine	I
bromine	Br	iridium	Ir
cadmium	Cd	iron	Fe
calcium	Ca	krypton	Kr
californium	Cf	lanthanum	La
carbon	C	lawrencium	Lr
cerium	Ce	lead	Pb
cesium	Cs	lithium	Li
chlorine	Cl	lutetium	lu
chromium	Cr	magnesium	Mg
cobalt	Co	manganese	Mn
copper	Cu	mendelevium	Md
curium	Cm	mercury	Hg
dysprosium	Dy	molybdenum	Mo
einsteinium	Es	neodymium	Nd
erbium	Er	neon	Ne
europium	Eu	neptunium	Np
fermium	Fm	nickel	Ni

Element	Symbol	Element	Symbol
niobium	Nb	strontium	Sr
nitrogen	N	sulfur	S
nobelium	No	tantalum	Ta
osmium	Os	technetium	Tc
oxygen	O	tellurium	Te
palladium	Pd	terbium	Tb
phosphorus	P	thallium	Tl
platinum	Pt	thorium	Th
plutonium	Pu	thulium	Tm
polonium	Po	tin	Sn
potassium	K	titanium	Ti
praseodymium	Pr	tungsten	
promethium	Pm	(wolfram)	W
protactinium	Pa	unnilennium	Une
radium	Ra	unnilhexium	Unh
radon	Kn	unniloctium	Uno
rhenium	Re	unnilpentium	Unp
rhodium	Rh	unnilquadium	Unq
rubidium	Kb	unnilseptium	Uns
ruthenium	Ru	uranium	U
samarium	Sm	vanadium	**V**
scandium	Sc	xenon	Xc
selenium	Se	ytterbium	Yb
silicon	Si	yttrium	Y
silver	Ag	zinc	Zn
sodium	Na	zirconium	Zr

Index

Acquired Immunodeficiency Syndrome. See AIDS.

Aduki Beans with Squash, 244

AIDS (Acquired Immunodeficiency Syndrome), 127

Air, quality of, 116

Akizuki, Tatsuichiro, 104, 144, 176

Alcohol, 103, 108-109, 133, 135

Alcoholism, 144

Allicin (in garlic), 210

Alpert, M.E., quoted, 99

Alpha radiation, 31, 33, 207

Aluminum, 32, 93, 116, 123-125, 133, 145, 226

Alzheimer's disease, 124, 133

American Cancer Society, 160, 168, 172

America's Toxic Protection Gap, 116

Americium-241, 42, 47

Ames, Bruce, quoted, 103

Anatomy of An Illness, 114

Anderson, K., quoted, 141

Anemia, 119-120, 123-124, 131

Antibodies, 110, 129-130

Antioxidants, 69, 154, 168

Arteriosclerosis, 152

Arthritis, 6, 69, 74, 119, 121, 126, 144, 152, 256.

Atherosclerosis. See Arteriosclerosis .

Atomic Age, the, 3, 25, 43, 70

Atomic Energy Commission (AEC), 5, 11, 21, 45, 53, 75-76, 87-88

Avoiding detrimental foods, principle of, 136

Background radiation, 23, 35-36, 38, 73, 77, 130

Bailey, Herbert, quoted, 104

Ball, Howard, 49

Barley, 105, 145, 156-157, 160, 164, 175, 228-230, 232, 244

B-complex vitamins, 119, 142, 164, 166, 158, 170, 178, 174, 186, 206

See also Folic acid; Pantothenic acid; Vitamin B1; Vitamin B2; Vitamin B6; Vitamin B12.

Bean Curry, 244

Beans, 107, 133, 147, 151, 153, 155-158, 161, 164, 167, 171-175, 183-184, 186-188, 201, 208, 226, 231, 234, 243-247

radioprotective factors in, 172

Becquerel, Antoine Henri, 34

Becquerel (Bq), 34

Bertell, Rosalie, 10, 27, 77, 89, 253

Beta radiation, 31

Bethe, Hans, quoted, 49

"Blocking agents, 154

Blood, 23, 30, 36, 38-40, 42, 69-70, 77, 98-99, 103, 105-110, 118, 121-123, 129-132, 134, 136, 139, 140, 142-144, 146, 148-149, 155, 160, 163, 165-166, 170, 174, 180, 183, 186-188, 203, 206, 207, 210, 256

acid-alkaline (pH) balance of, 107

protective properties of, 130-131

Bohr, Niels, 43

Book of Changes, 134

Book of Macrobiotics, The, 135

Broccoli, Sauteed, 240

Broccoli, Steamed, 241-242

Brodeur, Paul, quoted, 39

Bross, Irwin, 54, 74, 102

Brown rice, 156-157, 160, 164, 176-177

Brown Rice and Barley with Sunflower Seeds, 229

Brown Rice Salad, 230

Brown Rice with Wild Rice, 230

Buckwheat, 108, 157, 164, 233-234

Burkitt, Dennis, 157

Cabbage family, 119, 121, 123, 157, 167-169, 188, 201, 209

Cadmium, 44, 94, 107, 117, 122-123, 145, 208-209

Calcium, 25, 40, 68, 86, 92-93, 96, 98-99, 108, 117, 119, 122-124, 129-130, 132-133, 139, 141-143, 147, 150, 157, 159, 161, 163-164, 166, 170-171, 172, 174, 176, 179-185, 187, 205, 225

Caldicott, Helen, 17, 90, 96

Calories, 33, 102, 113, 137, 140, 143, 148, 151, 158, 160, 178

Cancer, 5-6, 17, 20-24, 28-29, 36-37, 41, 43, 45, 49-50, 54, 56, 60, 66-72, 76-77, 84, 88-90, 96, 100-105, 109-113, 122, 125-127, 134, 136, 141, 148-149, 153, 155, 160, 165-169, 172-173, 177, 186, 188, 203, 252, 256-257

colon, 103, 148, 151, 155

lung, 41, 43, 56

stomach, 103, 133, 153, 177

Cancer and Nutrition, 102

Cancer and the Philosophy of the Far East, 104, 134

Candida albicans, 144

Carbohydrates, 108, 113, 142, 162. See also Complex carbohydrates.

simple sugars vs. complex carbohydrates, 144, 151, 157, 160

Carotene, 167, 184, 186. See also Vitamin A.

Carrots and Onions, 236

Center for Defense Information, 59

Cesium-137, 45, 47, 61-62, 79, 96, 98, 100, 150, 167, 170, 174, 183, 187, 203-204

Chadwick, James, 43
Changing American Diet, The, 137
Chelation, 156
Chemicals. See Environmental contaminants.
Chernobyl, 5, 10, 43, 53, 83-85, 97, 149, 211, 257
Chick Pea Salad, Chilled, 245
Chiles, James, quoted, 44
Chili, Vegetarian, 246
Chlorine, 78, 94, 125, 127, 259
Cholesterol, 141, 155
Christensen, A. Sherman, 48
Chromium, 94, 145-146, 161, 163, 180, 183, 187, 205, 225
Cole Slaw, Picnic-Style, 239
Colitis, 124, 155
Complex carbohydrates, 144, 151, 157, 160
Concentration factor, 79
Corn, 56-57, 159-160, 164-165, 167, 170-171
"Corn-off-the-Cob" with Umeboshi, 237
Coronary disease. See Heart disease.
Cosmic rays, 32, 35, 258
Cousins, Norman, 114
Croquettes, 229 Crucifers. See Cabbage family.
Curie, Marie and Pierre, 34
Curie (Ci), 34
Cysteine, 158, 168, 178, 209, 225

Daikon radishes, 109
Daily meal planning guide, 186, 201

Dairy products, 136, 142 See also Milk.
Davis, Adelle, 168
Davison, K.L., 140
Defense Advanced Research Projects Agency, 49
Department of Agriculture, 133, 149
Department of Energy (DOE), 50, 54, 59, 64, 76
Department of Health and Human Services, 54
Diabetes, 6, 69, 74-75, 136, 140, 144, 146, 155, 161, 180, 187, 205, 256
Diet for a Small Planet, 162
Diet for the Atomic Age, 3, 18, 69, 113, 152, 154, 159, 226, 234, 236
Diverticulitis, 155
Doses. See Radiation doses.
Dr. Berger's Immune Power Diet, 160

Eat to Win, 160
Eggleston, David, 122
Einstein, Albert, 43
Eisenhower, Dwight D., quoted, 46
Electromagnetic spectrum, 29-30
Environmental contaminants, 114, 207
Environmental Protection Agency (EPA), 4, 39, 41, 61, 86, 116, 122, 124
Enzymes, 69, 110, 126, 155, 175, 207
Ershoff, B.H., quoted, 154
Fabrikant, Jacob, quoted, 101

Fat(s), 68, 108, 113, 117, 136-137, 140-141, 144, 148, 152, 154-155, 157-158, 160-161, 164, 167, 178, 184, 203, 206-207, 227

Fermi, Enrico, 43

Fiber, 139, 148, 154-159, 161-163, 165, 171-172, 174, 183-184, 188, 208

Binding ability of, 161

Fission, 19, 21-25, 43-45, 84, 96

Fluoride, 125-127

Fluoride: The Aging Factor, 125

Folic acid, 131, 166, 170, 186

Food and Drug Administration, 203

Food and Nutrition Board. See National Academy of Sciences.

Food chain, 8, 42, 78-79, 80, 82-84, 86, 116, 121, 124, 141, 148, 161-162, 178, 184, 204, 258

Food irradiation, 203-204

Foods

to avoid, 136, 149

chemicals in, 24-25, 115

fractionated, 136, 139, 145, 157, 162

neutral, 28, 68, 108, 133, 162-163, 179

protective, 8, 15, 28, 99, 120, 124, 129, 132, 140, 153, 155, 157-159, 167, 172, 177, 184, 201, 203-204, 252, 258

whole, 6, 16-17, 20, 28, 37, 48, 81, 87, 89-90, 137, 139, 141, 145, 151-155, 157-164, 172-173, 183, 186-188, 204, 208, 227-228, 233-235, 260

Free electrons, 67

Free radical scavengers, 69

Free radicals, 24-25, 66-69, 102, 111, 140, 146, 154, 161, 167, 169, 173, 184, 186-188, 203, 206, 209-211, 255

Fried Rice with Cabbage and Chick Peas, 231

Fried Tofu or Tempeh, 231

Fruit(s), 84-85, 103, 108, 133, 135, 137, 140, 143, 145, 149-150, 154-155, 159-160, 187-188, 201, 203

Gamma rays, 20-21, 32, 42, 203

Garlic tablets (Kyolic), 209, 225

Giron, Jose, quoted, 128

Gofman, John, 5-6, 10, 37, 45, 53, 101, 253

Grains, 81, 105, 107, 116, 123, 133, 137, 141, 145-147, 151, 153-155, 157-164, 167, 169, 171-172, 178, 183, 185-188, 201, 203, 208, 226-229, 232-234, 243

Cooking of, 149

Protective properties of, 157

refined, 123, 145, 162, 186

Gray (Gy), 34

Green Beans with Lemon Miso, 237

Green vegetables, 165-167, 170, 186-188, 208

Greens, Steamed, 242

Haas, Robert, quoted, 160

Health, effects of radiation on, 24, 69, 74-75, 136, 180, 252

Heidelberg Report, 78, 105, 114

Herb Book, The, 168

High blood pressure. See Hypertension.

Hijiki, 181, 183

Hiroshima, 16, 21-22, 73-75, 90, 101, 255
Hummus, 171
Hunt, Vilma, 41
Hydrochloric acid, 133
Hyperactivity, 119, 124, 144
Hypertension, 119, 122-123, 140, 144-145
Hypoglycemia, 144, 146, 187, 205

Immune system, 6, 17, 23, 39, 41, 68, 100, 102-103, 106, 109-114, 122-123, 124, 127-129, 140-141, 145-146, 154-155, 165, 186-187, 206, 208, 210
Infant mortality, 22-24, 72, 97, 99, 256
Insomia, 144, 205
International Atomic Energy Association, 84
Iodine, 23, 36-37, 45, 47, 91, 96-100, 183, 187
Iron, 97, 100, 131, 145, 157, 164-166, 176, 180, 184-185, 187, 205, 226 (chart), 46, 94, 98, 163, 170, 174, 181, 183, 225 (chart)
Isotopes, 29, 45, 55, 98-99, 257. See also Radionuclides

Jacobson, Michael, 137
Jensen, Bernard, quoted, 131
Johnson, Carl J., 59, 71
Justice Downwind, 49

Kahn, E.J., quoted, 159
Kennedy, Ann R., 172
Kepford, Chauncey, 43

Kidneys, 42, 106-108, 114, 117, 120, 123, 130, 133, 139, 147-148, 151, 164
Killing Our Own, 17, 48, 55, 76, 90
Kneale, George, 76
Kombu, 158, 183-184, 243-245
Kushi, Michio, 105, 135

Lappe, Frances Moore, quoted, 162
Lawless, Bill, quoted, 63
Lead, 32, 85, 94, 107, 117-122, 122-123, 127, 208-209
Lead in the Environment, 118
Lemoned Rice and Barley, 232, 244
Lemon-Miso Salad Dressing, 230
Lenihan, J., 99
Lentil, Salad, 245
Lentil Soup, 175, 245
Let's Get Well, 168
Leukemia, 21-23, 28, 36, 40, 49, 54, 69-71, 74-76, 89-90, 102, 105
Light, visible, 29
Little, John B., 172
Liver, 36, 42, 97, 106, 108-109, 112, 114, 117-119, 123-124, 139, 145-148, 151-152, 168, 203, 207
Lourou, M., 168
Low-Level Radiation Waste Policy Act of 1980, 62
Lust, John, quoted, 168
Lymph nodes, 42, 109-110, 114, 131
Lymphocytes, 109-110, 114, 130
Lysine, 171, 178

Magnesium, 93, 99, 108, 117, 119, 131, 140, 142, 163-166, 170, 180, 183-185, 187, 225

Mancuso, Thomas, 75
Marx, Jean, quoted, 112
Meat, 79, 81, 84-85, 107-109, 131, 133, 136, 146-149, 151-152, 154, 158-161, 163, 168, 176, 178, 184, 201, 206
Menu, sample, 202
Mercury, 94, 107, 117, 120-122, 209
Metals, toxic. See Toxic metals.
Methionine, 109, 158, 168, 178
Microwaves, 38-39
Milk, 21, 55, 60, 71, 83-86, 141-143, 150, 164, 166. See also Dairy products
Millet, 157, 159-160, 162-164
Millet with Crushed Almonds, 232
Minerals, essential, 98
Miso, 105, 109, 133, 153, 157-158, 174-178, 183, 186-188, 227
Muffins, Cornmeal, 235
Mustard Greens or Kale with Red Onions, 238
Mutations, 17, 20, 28, 252, 256-257

Nagasaki 1945, 104, 144
Najarian, Thomas, 77
National Academy of Sciences (Food and Nutrition Board), 103, 113, 118, 147, 150, 207, 255
National Cancer Center (of Japan), 177
National Council on Radiation Protection (NCRP), 35, 86, 97, 172
National Institute for Occupational Safety and Health (NIOSH), 38, 122

Non-ionizing radiation, 18, 29, 37-38
Nuclear Madness, 17, 90, 96
Nuclear medicine, 28, 36
Nuclear power plants, 4, 6, 9, 27, 36, 43-45, 49, 55, 60-62, 77, 188, 251-252, 255-256, 258, 261
Nuclear power plants, leaks in, 55
Nuclear Regulatory Commission (NRC), 4, 50-52, 90, 101
Nuclear Test Ban Treaty, 71
Nuclear Waste Policy Act of 1982, 62
Nuclear weapons, 10, 21, 24-25, 27, 36, 43-45, 48-50, 56, 59-60, 69, 71, 76, 256, 258
Nuclear Witnesses, 74, 77
Nuclear Workers, 75-77, 90
Nutrition Against Disease, 103
Nutrition Subcommittee of the U.S. Senate, 160
Nuts, 141, 153, 157, 159-160, 184-188, 201

Oatmeal, Breakfast, 228
Oats, 157, 164, 172
Obesity, 113, 140, 148
O'Connor, Egan, 37
Ohsawa, George, quoted, 104, 134
Oil, cooking with, 227
Optimal health, principle of, 136
Oriental Cabbage and Celery, 239
Osteoporosis, 144, 147
Ott, John, 38

Parkinson's disease, 124
Passwater, Richard, quoted, 98, 180
Pasta Variations, 234

Pasteur, Louis, 151, 210
Pasteurization, 142
Paul, T.M., 182
Pauling, Linus, 10, 70
Petkau, Abram, 23, 73
Pfeiffer, Carl, 104, 208
Physical Constitution and Food, 176
Phytates, 154, 157, 161, 163, 172-173, 183-184, 188
Plutonium, 19, 25, 37, 44-45, 48, 50-51, 55-56, 60, 62, 64, 78-79, 82, 84, 94, 97, 100, 127, 163, 170, 187, 259
Poisoned Power, 54-55
Pollard, Robert, quoted, 53
Pollution, 27, 65, 105, 112, 114-116, 123-125, 136, 139, 148, 161, 170, 206, 258.
See also Environmental Contaminants.
Polonium-210, 41-42
Potassium, 92, 96-98, 100, 132, 139, 150, 166, 170, 174, 180, 181, 183-185, 187, 204, 225
Poultry, 133, 136
Pritikin, Nathan, quoted, 162
Promethium-145, 34
Protease inhibitors, 172-173
Protein(s), 66, 68, 107, 111, 113, 131, 133, 137, 141-142, 147-148, 151, 157-161, 163, 165, 171-173, 176, 178-179, 184-185, 203, 206-207, 229, 243
Public Health Service, 254

Quickly Boiled Watercress and Carrots, 240

Radioactivity detectors, 42
measurements, 34, 82, 257
Radiation and Human Health, 44
Radiation doses, 91, 100, 105
Radiation Standards and Public Health, 54
Radio waves, 29
Radioecological Assessment of the Wyhl Nuclear Power Plant (Heidelberg Report), 82, 105
Radionuclides.   Radiolytic products, 203
Radionuclides, 29, 36-37, 45, 47, 61, 78, 82, 85-86, 91, 97
Radionuclides in Foods, 150
Radium, 20, 29, 34, 41, 61, 64
Radium-226, 34, 37, 42, 96
Radon, 40-42, 61
RBE (relative biological effectiveness), 33
Reagan, Ronald, 49
Recipes, 226-246
Recommended Daily Allowance (RDA), 141, 210
Rem (roentgen equivalent man), 233
Rice Porridge with Vegetables and Plum, 233
Roentgen, Wilhelm, 32
Roentgen (R), 32-34, 87
Rose, R.J., 140
Rosner, Frank, quoted, 128
Rouleau clumping, 38
Rutherford, Ernest, 43
Russell, Walter, 257, 259
Rye, 157, 164

Salad, Summer, 242
Salt, 144, 158, 179, 227

Sauces, 179, 227

Schizophrenia, 205

Sea vegetables, 133, 153, 157-158, 178-183, 186-188, 201,

Seasonings, 241

Secret Fallout, 17, 71, 73

Seeds, 141, 153-154, 157, 159, 161, 171-172, 188

Selective uptake, principle of, 7-8, 86, 92, 97, 147, 154, 159, 165

Selenium, 69, 94, 121-122, 161, 163, 188, 210, 225

Sell, J.L., 140

Sesame Salt, 236, 241

Shutdown: Nuclear Power On Trial, 75

Sievert (Sv), 35

Simone, Charles, quoted, 102

Skoryna, Stanley, quoted, 182

Sloyan, Patrick, quoted, 83

Smoke detectors, 42-43

Smoking, 41, 207

Sodium alginate, 119, 154, 158, 179, 181-183, 188

Solomon, Norman, quoted, 48, 55, 90

Sonnabend, Joseph, quoted,129

Spicy Sesame Noodles, 234

Spleen, 36, 109-110, 114

Squash Puree, Scrumptious, 241

Staffs of Life, The, 159

Sternglass, Ernest J., 17, 19-25, 72, 75, 129, 225, 257

Stewart, Alice, 22, 70, 74, 76, 100-101

Stokes, Charles, quoted, 51

Stress, 107, 112-114, 132-133, 146, 155

Strontium Metabolism, 99, 142, 182

Strontium-90, 23, 25, 30-31, 45, 47, 60, 71-72, 79, 86, 90, 93, 96, 98-100, 129-130, 132, 141-142, 150, 163, 170, 174, 181-183, 187

Sugar(s), 108-109, 133, 136-137, 143-146, 152, 154-155, 160, 163, 227

Sulfur, 94, 98, 167-168, 171, 209-210

Sulfur-containing amino acids, 109, 154, 158, 165, 188

Superoxide, 67

Supplements, 7, 112, 147, 166, 205-211, 208-212, 225

Susceptibility to radiation, 91, 107, 153, 175

Szilard, Leo, 43

Tahini, 185

Tamplin, Arthur, 53

Taylor, Alfred and Nell, 125-126

Television sets, 18, 36

Tempeh, 109, 153, 157-158, 174, 178-179, 186, 201

Three Mile Island, 15, 24, 51, 90, 97

Thymus, 23, 109-112, 114, 154, 161, 163, 165, 167, 170-171, 174, 183, 185-186, 188, 206-207

Tilden, J.H., quoted, 131

Tofu, 153, 157, 158, 161, 174, 178-179, 187, 201, 231, 236, 238-240

Toxemia, 131

Toxic metals, 117-127

Transfer factor, 79

Troll, Walter, 172

Ultraviolet rays, 35

Union of Concerned Scientists, 17, 52-53, 261

United Nations' Food and Agricultural Organization (FAO), 160

Uranium, 3, 9, 19, 37, 40-41, 43-44, 46, 56, 61-62, 64, 94, 107, 130, 251

VDT News, 38

Vegetable Stew, Old-Fashioned, 238

Vegetable Stock, 238, 246

Vegetables, 21, 81, 84-86, 103, 107, 109, 119, 121, 123, 133, 137, 140, 143, 147, 149-151, 153-155, 157-158, 160, 164-171, 177-183, 185-188, 201, 203, 208-209, 227, 236-242

Video display terminals (VDTs), 31-38

Vitamin A, 157, 167, 170-171, 174, 184, 186, 241

Vitamin B2, 145

Vitamin B6, 108, 112-113, 140, 145-146, 154, 161, 163, 167, 170, 185-186, 206-207, 209

Vitamin B12, 36, 97-98, 131, 158, 174, 176, 178, 183, 186

Vitamin C, 15, 108-109, 119, 122, 124, 126, 131, 157, 165-166, 170, 187, 205, 209, 225

Vitamin D, 118, 184

Vitamin E, 126, 140, 145-146, 154, 161, 163-164, 170, 184-185, 187

Wakame, 158, 183

Waldron-Edward, Deirdre, 182

War Resisters League, 59

Wasserman, Harvey, quoted, 17, 48, 55, 90

Waste, radioactive, 3, 12, 18, 27, 56, 59-64, 85, 115, 261, 263

Waste Paper, The, 64

Water, drinking, 24, 64, 78-79, 112, 120, 122-127, 256

Wattenberg, Lee, quoted, 169

Wheat, 116, 136, 145, 157, 159-160, 171, 208, 228, 233, 235

Williams, Roger, quoted, 103

Wyhl nuclear reactor, 78

X-rays, 20, 22, 24, 27, 29, 32, 35-37, 41, 70, 74, 87, 90, 166

X-Rays: Health Effects of Common Exams, 37

Yang, 134-135

Yavelow, J., quoted, 173

Yellow vegetables, 158, 167, 201

Yiamouyiannis, John, 125

Yin, 134-135

Zapping of America, The, 39

Zinc, 40, 94, 97-98, 100, 112-113, 119, 122-124, 127, 139, 145-146, 154, 157, 163, 171, 173-174, 180, 183-185, 188, 205-208, 225, 229

Zybicolin, 154, 174, 176, 183, 188